OPERATIONS RESEARCH-THE ART OF MAKING GOOD DECISIONS

Edited by **Kuodi Jian**

Operations Research - the Art of Making Good Decisions
http://dx.doi.org/10.5772/62933
Edited by Kuodi Jian

Contributors

Jernej Prisenk, Karmen Pažek, Črtomir Rozman, Matija Kaštelan, Martina Bavec, Susana Garrido Azevedo, Adolfo Minjarez-Sosa, Fernando Luque-Vásquez, Milos Radomir Vasic, Zagorka Radojevic, Robert Rekecki, Jan Derco, Rastislav Kuffa, Jozef Dudáš, Barbora Urminská, Jana Marová, Wolfgang Ossadnik, Fabian Burrey, Thorben Hustedt, Kuodi Jian

Published by InTech

Janeza Trdine 9, 51000 Rijeka, Croatia

Publishing Process Manager Martina Usljebrka
Technical Editor SPi Global
Cover InTech Design team

First published January, 2017
Printed in Croatia

Additional hard copies can be obtained from orders@intechopen.com

Operations Research - the Art of Making Good Decisions , Edited by Kuodi Jian
p. cm.
Print ISBN 978-953-51-2817-5
Online ISBN 978-953-51-2818-2

For EU product safety concerns:
IN TECH d.o.o., Prolaz Marije Krucifikse Kozulić 3, 51000 Rijeka, Croatia,
info@intechopen.com or visit our website at intechopen.com.

Meet the editor

Dr. Kuodi Jian is a noted scholar in the field of computer science. He holds a BS degree in Computer Science from the University of Mary Hardin-Baylor in Texas, USA, and an MS degree in Computer Science and a PhD degree in Computer Science and Operations Research from the North Dakota State University in Fargo, USA. He worked as a computer system architect at Banner Health System, Fargo, North Dakota, as an associate professor in Metropolitan State University (MSU) since 2003, and he currently takes the role of ICS Graduate Director in MSU. Dr. Kuodi Jian is active in his research. He published a book *A Graph Planning Procedure within an Agent Architecture Fast Planning and Distributed Agent Architecture*, book chapters "Knowledge Management in Bio-Information Systems" and "Introductory Chapter: Real-Time Systems," and numerous journal/conference articles in the areas of algorithms, programming languages, real-time operating systems, operations research, database systems, web service-oriented architecture (SOA), artificial intelligence, computer hardware, and computer simulation.

Contents

Preface

In a nutshell, operations research (sometimes called operational research) is a decision science. It studies the subject of how to make the best decision given the constraints at hand. When applying to different fields, it manifests itself as many forms: in military, it takes the form of using strategies to win a war; in business, it takes the form of making most profit; and in mathematical modeling, it takes the form of finding optimal solution by using derivative, or simplex algorithm of linear programming.

Operations research is an important topic. The importance of winning a war, making maximal amount of profit, and building a correct math modeling is obvious. Here, I want to point out another aspect of operations research that is often overlooked and vitally important to us: efficient use of resources. Currently, human species are consuming natural resources at an astonishing rate: resources such as crude oil, drinkable water, safe food are in danger or soon will be exhausted. We must be aware and take actions before it's too late. One of the solutions is to make efficient use of natural resources, and operations research offers such a tool. In this sense, the topic of the book is important and relevant to everyone.

The content of operations research discussed in this book covers a wide range of areas and has some unique features:

- Both theory and practice are presented. The editor of the book understands the importance of both theory and practice and is careful when selecting chapters. As a result, the book has a good balance of decision theory and application practices.
- The content covers a wide spectrum of application areas. As you can see, the topics range from mathematical modeling to performance measurement and from industrial process to business case study. But they all contribute to the theme of making good decision.
- The book reflects the rich background of chapter authors and a wide variety of viewpoints. Authors of this book come from different countries and different walks of life. This fact contributes to the attractiveness and the richness of the book.

It's exhilarating to know that this book is the result of contributions by practitioners, researchers, scientists, and scholars from many countries, like Slovak Republic, Slovenia, Serbia, Mexico, Portugal, Germany, and the USA. This book will be useful to a wide range of audiences: university students/professors, government policy makers, engineers, and businessmen who are interested in operations research.

Dr. Kuodi Jian
Metropolitan State University, Computer Science Faculty
Department of Information and Computer Sciences
Minnesota, The United States of America

Chapter 1

Introductory Chapter: Operations Research

Kuodi Jian

Additional information is available at the end of the chapter

http://dx.doi.org/10.5772/66835

1. Introduction

Operations research (sometimes referred to as management science or decision science) is a subject that deals with the art of making good decisions under specified constraints. In real life, we are often faced with complex situations for which no simple answer can be found. Thus, the topic of making good decision (operations research) is both intriguing and relevant. Most people consider operations research as a subfield of mathematics. As pointed out below, the criteria of good or bad decisions are often affected by culture, viewpoints, and other factors.

Other than moral issues, there are **two important keys** to make good decisions: **good information** and **the skill of making good decisions** based on the information at hand. By "good information," we mean that all essential factors are captured (tools used to obtain quality information could be verification and validation); by skilled decision making, we mean the application of appropriate solutions to different problems (e.g., to solve well-formed mathematical optimal problems, we use calculus or the simplex algorithm; to solve poorly defined problems, we use empirical trial-and-error methods or ad hoc methods). The subject of operations research covers both the acquiring of "good information" and "the skill of making good decisions." **Figure 1** shows the relationship among these entities.

Depending on the problem domain, decision-making skills can be regarded as either a science or an art. When a problem is a well-defined mathematic problem, we are able to use scientific methods such as calculus, linear programming, integer programming, dynamic programming, and simplex algorithm; on the other hand, when a problem is poorly defined, we can only use trial-and-error methods, heuristic methods, or ad hoc methods. Since trial-and-error methods and ad hoc methods are non-repeatable, they are regarded as art. Making a good decision under multiple constraints (especially where different criteria are involved) is never easy; the topic of operations research is not easy and has a lot of variations.

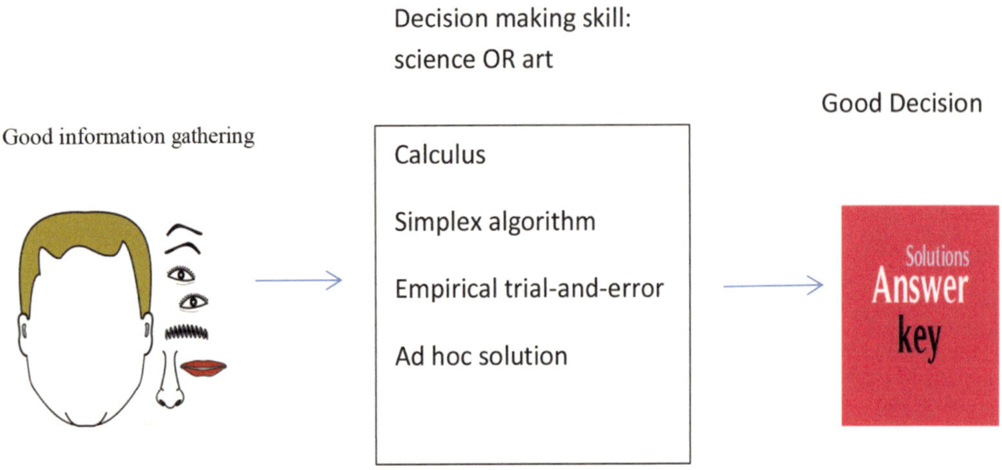

Figure 1. The relationship among obtaining information, decision process, and tools.

In this book, we cover a wide range of applications that employ a variety of decision-making skills (including problems that are poorly defined). Therefore we subtitled the book as "an art of making good decisions."

Since decision making is an essential part of our life, the application of operations research covers a wide range of areas such as military, business, mathematics, and resource allocation among competing parties, just to name a few.

When focusing on decision making, operations research has a long history.

2. History

Chinese people used good decision-making strategies to military more than 2000 years ago. In the book "Master Sun's Art of War" by Master Sun (also called Sun Tzu), important aspects of warfare are summarized into 13 chapters. The essence of the book is the strategies of how to motivate soldiers and leverage tactical advantages (win the battle of wits). The first chapter of the book talks about "Detail Assessment and Planning," and the last chapter of the book talks about "Intelligence and Espionage." From the available evidence, we know that the book was completed sometime between 500 and 450 BC. Some of the well-known strategies are as follows [1]:

> 故曰:知彼知己, 百戰不殆;不知彼而知己, 一勝一負;不知彼, 不知己, 每戰必殆。
>
> *So it is said that if you know your enemies and know yourself, you will not be put at risk even in a hundred battles.*
>
> *If you only know yourself, but not your opponent, you may win or may lose.*

If you know neither yourself nor your enemy, you will always endanger yourself.

This has been more tersely interpreted and condensed into the Chinese modern proverb:

知己知彼，百戰不殆。 *(Zhī jǐ zhī bǐ, bǎi zhàn bù dài.)*

If you know both yourself and your enemy, you can win numerous (literally, "a hundred") battles without jeopardy.

Other two high-order thinking of military campaign contained in that book are the following [2]:

"The supreme art of war is to subdue the enemy without fighting."—**Sun Tzu, The Art of War**

"If your enemy is secure at all points, be prepared for him. If he is in superior strength, evade him. If your opponent is temperamental, seek to irritate him. Pretend to be weak, that he may grow arrogant. If he is taking his ease, give him no rest. If his forces are united, separate them. If sovereign and subject are in accord, put division between them. Attack him where he is unprepared, appear where you are not expected."—**Sun Tzu, The Art of War**

The effectiveness of Master Sun's idea is well tested. At his time, Master Sun helped the warlord of his state Wu to defeat the much stronger enemy state Chu. Even today, many people benefit from the strategies of Master Sun. The following episode is taken from the History website and illustrates the point [3]:

Ever since The Art of War was published, military leaders have been following its advice. In the twentieth century, the Communist leader Mao Zedong said that the lessons he learned from The Art of War helped him defeat Chiang Kai-Shek's Nationalist forces during the Chinese Civil War. Other recent devotees of Sun Tzu's work include Viet Minh commanders Vo Nguyen Giap and Ho Chi Minh and American Gulf War generals Norman Schwarzkopf and Colin Powell.

Meanwhile, executives and lawyers use the teachings of The Art of War to get the upper hand in negotiations and to win trials. Business-school professors assign the book to their students and sports coaches use it to win games. It has even been the subject of a self-help dating guide. Plainly, this 2,500-year-old book still resonates with a 21st-century audience.

People around the world have been using optimization for a long time. For example, the sizes of some jade items excavated from Han dynasty's bury sites in China are optimal in terms of their surface areas vs. their weights. Ancient Egyptians built remarkable structures called pyramids as shown in **Figure 2**.

These types of structures were built with the correct proportion and angle, which is 52.606° for their top angle. When a pyramid is built that way, it will preserve certain energy. Some interesting aspects of pyramids discussed by the website are as follows [5]:

Some of the effects are:

Food kept under the pyramid will stay fresh for two to three times longer than uncovered food. Artificial flavorings in food will lose their taste, but natural flavors are enhanced.

Figure 2. A picture of pyramids [4].

The taste of foods change; they become less bitter and acidic.

When we take a spectrographic reading of the treated item, it will show a change in the molecular structure.

The pyramid will dehydrate and mummify things, but it will not permit decay or mold to grow on them.

There is also a slowing or complete stopping of the growth of microorganisms.

Kirlian photographs of human subjects show the aura to be significantly brighter after a 15-min exposure period.

Pyramid research:

Bill Kerell has been a pyramid researcher for about 17 years. He has performed many experiments using brine shrimp. Brine shrimp usually live for 6 to 7 weeks, but under the pyramids, Bill observed that brine shrimp can survive for over a year. He also noticed that pyramid-grown shrimp grew two to three times larger than the normal ones. Bill has also conducted a lot of researches with humans.

Bill and his associates also have found that hypertensive individuals become tranquilized, but lethargic people become energetic again.

All these show that humans have been using the knowledge of optimization for a long time and some of their creations are still not fully understood and still have research value for us.

However, "the first formal activities of Operations Research (OR) were initiated in England during World War II, when a team of British scientists set out to make scientifically based decisions regarding the best utilization of war materiel. After the war, the ideas advanced in military operations were adapted to improve efficiency and productivity in the civilian sector" [6].

3. What is a good decision?

In this section, we will answer the question "what is a good decision?" At a first look, it appears simple. But, when taking a closer look, you will find that the answer is not so simple. First, before you are able to answer the question, you have to understand the concept of decision criteria. Second, you need to understand the dynamics of requirements, relationships between natural laws and decision criteria, the point of views, and the culture. Third, when moral is involved, there is no simple binary answer, and instead, it becomes a philosophical answer.

In decision science, a criterion is a reference yardstick against which the quality of a decision is measured. If the criterion is met, the decision is good; otherwise, the decision is bad. Now, you may wonder: how decision criteria are made.

3.1. Relationship between natural laws and decision criteria

Whether you admit it or not, in this universe, there are a lot of natural laws in existence and these laws affect our decision-making processes. The effects of natural laws manifest themselves by rewarding those decisions conforming to the laws and punishing those decisions violating them. Throughout the years, people create decision criteria with natural laws in consideration. In fact, natural laws determine and affect decision criteria. For example, when standing on the surface of the earth, we will feel something pulling us down and we call this non-visible force "gravity." The law of gravity helps to produce decision criteria that avoid fall but favor safety.

One directional passing of time makes death irreversible. This will produce decision criteria that favor life but recoil from death.

Another example is the conservation law. In a closed system, we cannot create something from nothing nor can we destroy something and make it disappear; we can only change its existing form from one to another. Criteria nurtured by this law are the human disposition toward resource saving and refraining from wasting. When applying criteria derived from the conservation law, we position ourselves to problems that exhibit themselves as optimal problems. This is why most people think operations research is equal to optimality finding. As you can see, in reality, operations research (or decision science) is more than that.

One natural phenomenon that interests us is the randomness. This phenomenon expresses itself in different ways: when generating a random number, tossing an unbiased die, or flipping a fair coin. It manifests itself as if there were a designer of this law in the universe who is fair. Anyone with a clear mind, regardless whether he or she believes there is a designer or not, would have faith that the number of heads he or she would get when flipping a fair coin is approximately half of the total tosses, given that the number of try is large. It is this subconscious belief that affects our statistic criteria.

3.2. Decision criteria are affected by cultures and view points

Decision criteria are affected by cultures. By culture, we mean the norms of a community and its sanctioned customs. As we are brought up in a community, we carry the fingerprints of that community. These fingerprints will be reflected in our decision criteria. For example, most of Chinese will not donate their body parts after their death. The mainstream Chinese culture values whole body (the belief of reincarnation and a perfect body in the next life cycle). As a result, you will hardly see a Chinese who selects organ donation on the back of his or her driver's license.

In terms of viewpoints, I will use the following true story to illustrate their effects on decision criteria:

In ancient China, there was a period called "Spring and Autumn" (771-476 BC). During that time, there were two opposing states Wu (the state king is named Fuchai) and Yue (the state king is named Goujian). The king of Yue was captured and humiliated by the king of Wu. After returning to his state, Goujian (the king of Yue) vowed revenge. One of his high-ranking officers named Weng Zhong conceived 10 strategies to weaken and destroy the enemy state Wu. In fact, he only used 3 of his 10 strategies. By the end of his third strategy, the enemy state Wu was destroyed and the king Fuchai was captured and killed.

Now, let's take a look at one of Weng Zhong's strategies: causing famine in Wu. In one autumn season, Weng Zhong picked tons of high-quality un-threshing rice (rice with skin so they can be used as seeds for next year) freshly out of the rice fields. He secretly steamed them and made them inert. Then, he gave these rice to the enemy king Fuchai as a token of obedience. Sure enough, Fuchai took the bait and asked the farmers of his state to use these high-quality rice as the seeds for next year's crop. Thus, a big famine ensued in Wu the following year.

To judge Weng Zhong's famine-causing strategy, we will have different results depending on how you look at it. It would be an effective and good decision if you are from a view point that wants Wu to be destroyed; on the other hand, it would be a bad one if you are judging it from the conservation law's point of view or from Fuchai's point of view.

3.3. Decision criteria are affected by the requirements

When talking about good decisions, we need to be aware of the requirements attached. In many problems, the same objective variable will take different optimal values when the requirements are different.

For example, let us assume that we have a wire with length of 10 inches. and we ask the following two questions (see **Figure 3**):

1. What is the largest area that can be formed by using this wire (given that you can use any two-dimensional shapes)?

2. What is the largest area that can be formed by using this wire (given that you can use any different rectangular (including square) shapes)?

Answer 1: From the knowledge of algebra, we know that the circle will give us the maximum area. The problem boils down to solving the following two equations:
$D\pi = 10$
$\text{Area} = \pi D^2/4$
Thus, we get area = $\pi D^2/4 = [\pi(10/\pi)^2]/4 = 100/4\pi = 7.96$ inch^2

Answer 2: From the knowledge of algebra, we know that the square will give us the maximum area given that the shape must be rectangle. The problem boils down to solving the following two equations:
$4b = 10$
$\text{Area} = b^2$
Thus, we get area = $b^2 = (10/4)^2 = 6.25$ inch^2

As you can see, these two problems have the same optimization objective variables, but we get different results because of the different requirements. Thus, we conclude that the requirements in a decision problem play important roles.

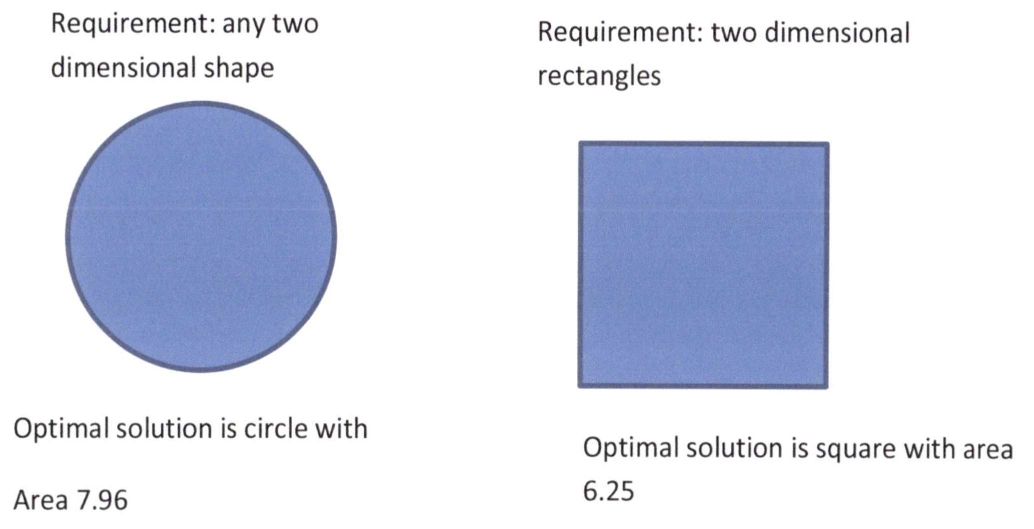

Figure 3. The effects of requirements.

3.4. Decision criteria are affected by moral beliefs

People are often faced with situations in which there are no right or wrong answers. We usually call these moral dilemmas. The criteria used here are mainly affected by a person's moral beliefs. For example, when a doctor is treating a terminally ill patient suffering from pain, the decision of whether to prescribe pain-relieving drugs such as Marijuana or morphine is affected by his/her belief system. Depending on how strongly he/she feels (pain vs. controlled drugs), the doctor acts accordingly to what he/she thinks appropriate (meeting his/her decision criteria).

Allowing a terminally ill patient to die at his/her will is also a decision affected by moral beliefs. When morals are involved, decision criteria are complicated since we are carrying the whole baggage of our belief systems. A lot of times, there are no simple answers (we only see and pick aspects that make us comfortable—beauty is in the eye of the beholder).

4. Overview of the chapters

In this book, we have carefully selected a set of manuscripts that are written by authors from different backgrounds. The selected articles have a broad spectrum of topics ranging from theory to application. On the other hand, all the topics are centered on the main theme of making good decisions. Thus, readers get the benefit of wide exposure to the ins and outs related to the subject. In the following, I will give you a brief introduction to each of the remaining chapters.

Chapter 2 "Improving Informational Bases of Performance Measurement with Grey Relation Analysis" written by Thorben Hustedt, Ossadnik Wolfang, and Burrey Fabian. The main contribution of the chapter is the presentation that provides a partial view on Grey Systems Theory (GST) as a conception to improve poor data situations for Performance Measurement (PM) and to operate with a few data already at hand. The chapter gives not only concepts related to GST, GST's element called Grey Relation Analysis (GRA), Performance Measurement (PM), and Key Performance Indicators (KPIs) but also an example of applying GRA analysis to a PM problem.

Chapter 3 "Application of Lean Methodologies in a Neurosurgery High Dependency Unit" written by Ricardo Balau Esteves, Susana Garriddo Azevedo, and Francisco Proenca Brojo. The main contribution of the chapter is the application of **Lean methodologies** to a Neurosurgery High Dependency Unit (NHDU). The manuscript presents the results; shows the research results; does the statistical analysis on the results; and points out the benefits of applying lean methodologies. The research method used is "an action research supported by a longitudinal mixed method approach with a one-group within-subjects pretest-posttest experimental type."

Chapter 4 "Iteration Algorithms in Markov Decision Processes with State-Action-Dependent Discount Factors and Unbounded Costs," written by Fernando Luque-Vasquez, and J. Adolfo MinJarez-Sosa. The main contribution of the chapter is the study of control models with state-action-dependent discount factors, focusing mainly on introducing approximation algorithms for the optimal value function (value iteration and policy iteration).

Chapter 5 "Mathematical Modeling of Isothermal Drying and Its Potential Application in the Design of the Industrial Drying Regimes of Clay Products," written by Milos Vasic, Zagorka Radojevic, and Robert Rekecki. The main contribution of the chapter is the creation of a link between the comprehensive theory of moisture migration during drying and the setup of the non-isothermal drying process.

Chapter 6 "Financial Feasibility Analysis of Natura Rab Business—Case Study," written by Karmen Pazek, Matija Kastelan, Martina Bavec, Crtomir Rozman, and Jernej Prisenk. The main contribution of the chapter is the case study on the economic validity of the three projects related to a family firm called Natura Rab and the model developed using Microsoft Excel on Net Present Value assessment. The model could be used in other similar projects and investment to support decision making.

Chapter 2 "Influence of Phosphorous Precipitation on Wastewater Treatment Process," written by Jan Derco, Rastislav Kuffa, Barbora Urminska, Jozef Dudas, and Jana Kusnierova. The main contribution of the chapter is the contribution of the methodologies for decision making related to finding most effective ways of removing phosphorus from wastewater, related to the selection of precipitating agents such as Fe^{2+}, Fe^{3+}, and Al^{3+} salts on sedimentation and thickening characteristics of sludge in wastewater treatment.

The above chapters cover a wide range of topics that are centered on the theme of operations research (decision science). We wish you enjoy reading rest of the book.

Author details

Kuodi Jian

Address all correspondence to: kuodi.jian@metrostate.edu

Metropolitan State University, Saint Paul, MN, USA

References

[1] Internet source: Wikipedia. Internet source: https://en.wikipedia.org/wiki/The_Art_of_War. Retrieved on 10/31/2016.

[2] Internet source: Goodreads. Internet source: https://www.goodreads.com/work/quotes/3200649. Retrieved on 10/31/2016.

[3] Internet source: History. Internet source. http://www.history.com/topics/the-art-of-war. Retrieved on 10/31/2016.

[4] Internet source: Wallike. Internet resource: wallike.com, retrieved 3/12/2014.

[5] Internet source: Algonet. Internet source: http://www.algonet.se/~anki-p/pyramid-power.html. Retrieved 3/8/2014.

[6] Taha, H. (2011). Operations research an introduction, 9th edition. Prentice Hall, Inc., Upper Saddle River, New Jersey, USA.

Chapter 2

Improving Informational Bases of Performance Measurement with Grey Relation Analysis

Thorben Hustedt, Wolfgang Ossadnik and
Fabian Burrey

Additional information is available at the end of the chapter

http://dx.doi.org/10.5772/65286

Abstract

Performance measurement (PM) needs objective empirical data with causal relevance in order to steer and control financial performance generation. In business practice, there is often a lack of such objective data. A surrogate might be collected subjectively based on data generated by questioning corporate experts. Such an involvement of subjects can rapidly lead to an immense extent of data that (partially) imply incomplete information. To handle this imperfection of data, the Grey systems theory (GST) and especially its element, the Grey relation analysis (GRA), seem to be methodologies able to improve informational bases for PM purposes. Therefore, GRA is able to reveal those performance indicators that considerably influence the corporate financial performance, the key performance indicators. GRA is able to supply valid results with only four data points of a time series. Hence, the GST provides an improvement of the PM framework in situations of incomplete information, which is demonstrated in the following.

Keywords: small samples, performance measurement, performance indicator selection, causal ambition

1. Introduction

In business practice, empirical data with causal relevance for financial performance generation are required for steering and controlling demands. Often there is a shortage of such data. Therefore, a severe problem has to be solved by the management. From the development and implementation of measurement and management systems, for example, performance management and measurement, a provision of causal-oriented data as a quantitative basis for steering and controlling purposes can be expected. PM as the quantitative database of management control operates as an information supply system for the performance management. The current relevance of the topic is shown by Rigby and Bilodeau [1] who expose the balanced scorecard (BSC) as one of the most popular management tools for strategically oriented performance management. A comprehensive BSC also requires the identification of causal interdependencies between indicators that drive the corporate's financial performance. But in business practice, often a shortage of especially objective data exists. To derive causal hypotheses on financial performance generation, subjectively based data can be collected and used in the framework of a PM design as a surrogate. Afterwards, these data can be intersubjectivated groupwise and objectivated by statistical validation.

Generating subjectively based data by interviews or questionnaires often leads to an excess of such data implying problems of handling. In this case, appropriate performance indicators have to be selected which contribute to the success of an organization. Identifying and ensuring an effective PM demands a focus on the cause-and-effect relations between these performance indicators. Their estimation and validation induce methodical questions how to cope with imperfect information. This challenge *inter alia* demands analytical decision support. Besides the known disciplines and methods to handle imperfect information, like stochastics, Fuzzy Mathematics or DEMATEL (as a technique of groupwise intersubjectivation), this chapter provides a partial view on Grey systems theory (GST) as a conception to improve poor data situations for PM and to operate already with a few data.

As organizations often do not have sufficient objective databases for PM purposes, they must refer to subjective data usually filtered out of tacit knowledge stemming from employee interviews. These finally lead to a large number of performance indicators determined by the means and relations of the fixed corporate strategy and its usage of identified cause-and-effect relations which is indispensable for causally ambitioned performance control. This is the framework that demands an evaluation and reduction of the obtained variety of indicators to main key performance indicators (KPIs).

For instance, 50 performance indicators are denominated as candidates by a company's employees with expert status. Hypostatizing the causal interdependencies of these would leads to a challenge without any operations research (OR) support. In addition, how could an organization obtain quick but also valid information for the selection of the KPIs, without multicriteria decision support, if the Statistics require a sample size implying longstanding data collection?

Do there exist methods to transform subjectively based data into intersubjectivated ones reaching closer to quasi-objective data and therefore allowing more detailed conclusions for the PM context?

Such methodology is made available by the GST that has been developed to handle situations with incomplete information that cannot be coped by other support disciplines. Thus, performance indicators can be selected by aid of the Grey relation analysis (GRA) based on subjective information. GRA analyses the geometric relationships of compared discrete objectives as well as of subjective indicators and is able to operate with a sequence length of minimally four data points. In situations with databases being too small for statistical analyses, processes of intersubjectivation or validation become possible. With GRA, PM would be enabled to prepare an order of the KPI priorities resulting from the geometrical similarity of the performance indicators' time series to the sequence of the top strategic financial performance ratio. In addition, it is also possible to display the interdependencies between the residual indicators in a network or in a causally ambitioned map by GRA to steer and control the performance generation in the PM system context.

2. Performance management and performance measurement

To focus the whole company on a long-term financial success, it is necessary to reflect and if required to recombine the objectives of the corporate strategy on every single company level, in each business unit and in the cognitive systems of the employees. Thus, the integration and therefore the implementation of the corporate strategy ensures the value creation in an organization. This value creation is also known as the generation of financial performance. The term "performance" is much discussed and underlies no standardized definition. It only becomes clear by an individual corporate-specific description [2]. The special task of the PM is to provide an information supply system for the management by finding the causal relationships that are related to financial performance. The causes of financial performance are not only financially dimensioned. The challenge of a thriving business is to include nonfinancial performance measures often anchored in the intuitive implicit knowledge of the employees. The ability to respond to altered circumstances presupposes an update of critical success factors [3, 4]. An entire focus of an organization on a backward-looking financial performance indicator system as it was usual in traditional Management Accounting with its reference to the decomposed structure of financial ratios (e.g. DuPont scheme) is inconceivable in today's dynamic business and Management Science. Instead, a new operational framework is necessary. Such a scope should include all relevant aspects of the corporate performance [4, 5]. This superior framework is tailored as a management and control system and is also known as the Performance Management of the organization. To provide an adequate information basis for the Performance Management, a measurement of the KPIs is necessary [6]. Therefore, the PM addresses three central functions: measurability of financial and especially nonfinancial indicators, identification and selection of the most important indicators that drive financial performance and lead to value generation. Thus, an additional transparency for the different members of the organization is provided. For this, a sound knowledge of employees about the process of financial performance generation is necessary [7].

Hence, the interaction of Performance Management and PM shows that these two internal corporate systems cannot be separated. The PM may be viewed double-edged: First, as a feedback-oriented system that supplies Performance Management with norms and information on current processes by data measured in the past and presence. Thus, a base to derive counteractions exists. Furthermore, a feedforward tool is made available, which informs about failings of the conceptual framework so that a new causal model will have to be developed, validated, and implemented [4]. Without any knowledge about the interaction between those systems, the organization misses the opportunity to control and dominate the performance-generating process [2, 8].

The performance-generation process has multidimensional aspects incorporated by the responsibility of multiple causes that lead to an unidimensional financial effect specified by the owners of an organization [4, 9]. Consequently, the interaction of the multidimensional PM and the Performance Management conduce to the improvement of the corporate performance. For quantifying the financial and nonfinancial measures, the PM serves support for a performance recording. Often a shortage of available, objective empirical data for the representation of performance indicators occurs, which has to be handled by the management control. In case of missing objective data, it is indispensable that the PM manages this problem by collecting subjectively based data on the basis of surveys or interviews which enable a quasi-objectivation of these measures [10, 11]. Even if organizations should have historical objective data, subjectively based data should not be ignored. Many times, historical data have been collected in varying frequencies and ranges. In this case, an usage within a PM seems to be inappropriate [12].

Revelations of the interdependencies between the KPIs that are essential for the value creation or rather the performance improvement can only be determined by sufficiently articulated knowledge. At this, it is necessary to differentiate between the explicit knowledge on the one hand which is simple to communicate and can be made available to all individuals that want to use it. On the other hand, there is the intuitive implicit knowledge that makes important performance-related causal relationships available [13, 14]. The implicit knowledge is characterized by four conditions: difficult to imitate, hardly to replace, only transmittable to a limited extent (not by the normal use of language), and scarce existence [13]. The tacit knowledge is, however, very difficult to create because it has been sharpened over years in extensive activities and experience of individuals. To evoke this dormant, subject-bound, intuitive knowledge, Abernethy et al. [10] propose to interview or rather execute subjective questionnaires so that the employees give partly insights into their tacit knowledge. This results in a variety of subjectively based data which first need to be reduced to a manageable level and can be intersubjectivated to work with. Here, the task of the PM should be based on an adequate—even optimal—complexity reduction [15]. Thus, the immense amount of subjective data has to be channelled, properly. Besides the PM has to concentrate on the essential factors with the aid of intersubjectivated data. All this is taking place to avoid that the PM System is more confusing than helpful.

2.1. Strategic alignment of performance indicators

The PM should not only be understood according to the phase of validating the established hypotheses at the beginning of the PM process but rather by Bourne et al. [16] as a tool to identify appropriate indicators covering structure and processes of an organization in a dynamically changing environment. To focus on inadequate measures would constitute a resource-wasting framework. Hence, the organizational, multidimensional PM System requires a selection of such KPIs endogenously linked to the corporate strategy and thus able to improve the performance [17]. Various studies [18, 19] detected that systems that are constructed as too complex have a negative influence on the performance. Too complex systems lead to an overload of information and consequently cause an increase of administrative costs [20]. Therefore, the amount of KPIs has to be limited to a level cognitively manageable by the members of the organization [17, 21].

On account of a lack of objective data, organizations may refer to subjective estimations stemming from samples being too small or too fragmented to apply statistical methods (**Figure 1**). Small or fragmented data sizes lead to incomplete information. This problem is to be solved by the GST. Fuzzy Mathematics, which focus on experience data of an individual, are characterized by a clear content (intension) but by unclear (not determined) quantitative boundaries of an expression—for example "very strong"—(extension of information). GST is more suitable with concepts of multiple meanings (e.g., performance), is additionally able to handle fuzziness situations and disposes of a clearly defined extension [22]. Thus, the above-mentioned problem of poor and incomplete information is almost impossible to solve with Fuzzy Mathematics or Statistics. The incomplete nature of the information needs to be managed in the PM context. A subjective query that was collected over a small number of periods can be considered as an incomplete information, if the experts of the organization deliver only a few estimations of the extent of an indicator [23]. Reducing the volume of performance indicators needs subjectively-based and thus poor information [24]. The organizational challenge is to solve this problem by providing valid results for PM also in case of small samples in situations of incomplete information. This would be possible by reference to support models for comprehending and decoding the problems of the system [25, 26].

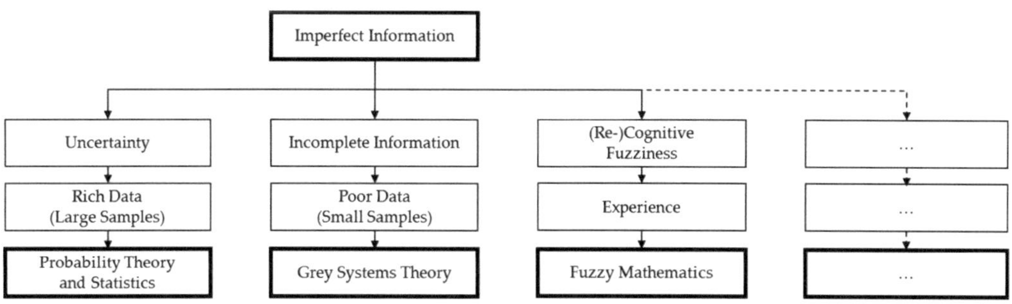

Figure 1. Imperfect information, situations and instruments.

In the PM context, it is important that a strategy is formulated as simple as possible [27]. If the extension of the strategy is then reduced to a manageable minimum, the organization possesses a list of factors most important for performance generation [28]. But this does not deliver a sufficient condition to control an organization successfully. Instead, it is essential to know how the factors are interrelated to actuate the right "lever" for an increase in financial performance [29]. Therefore, a causally ambitioned network of interdependencies of the KPIs seems to be useful [30].

2.2. Causal mapping

The performance of an organization can be interpreted as a result of past actions of the managers. To explain this performance, a causally ambitioned model with all relevant relations between the considered indicators is indispensable. Thus, the process leading to performance can be visualized. Such an illustration (e.g., a map) delivers – especially if structured—a blueprint for implementing the corporate strategy [31].

A map generally provides the visualization of a reference framework. In the 1970s, the political scientist Axelrod [32] spread the methodology of cognitive maps that should illustrate simplified social studies. A cognitive map provides an optical representation of the structures people perceive in their environment [9]. Cognitive maps serve management with a tool for evaluating alternative business situations in order to meet better an uncertain, dynamic corporate environment and to simplify complex issues [33]. Here, the organization should, however, focus on a visualization of tacit knowledge [10].

A simple list of the most important corporate strategy factors would point out the indicators the organization has to focus on. As an enumeration, such a list, however, would not represent the interdependencies within the system. To control as well as to monitor the performance generation, it is necessary to understand the causal relations between the KPIs [34, 35]. Therefore, it is fundamental to keep in view the cost–benefit ratio: a too detailed map costs a large amount of time [10], a graphical apposition of ovals does quantify dependencies in the system [32]. GRA as an OR management support is simple in usage and provides meaningful results already after a few periods. Additionally, it even enables a visualization of the outcome within a relational network [36].

In contrast to parametric approaches like the correlation analysis, nonparametric mapping approaches are much more able to represent the multidimensionality of the performance generation. By avoiding assumptions, nonparametric approaches focus on mapped causal relationships among the measures based on their perceived environment [11]. Organizations tend to skip a statistical validation of their causal model. The reasons for this are the perceived obviousness of the model, the time exposure or rather the high validation costs [20, 37, 38]. The changing dynamic and competitive environment requires an adjustment of an organization's causal model to adapt the strategy continuously. In order to meet this condition sufficiently, an ongoing customization of an organization's cause-and-effect network is not manageable with regard to time and costs that appear by longstanding serial questionnaires [39].

A sole focus on subjectively based data can lead to systematic judgment errors by incorrect estimations of individuals. Thus, such data are to be considered as incomplete because of small or fragmented sample sizes [39]. In addition, subjectively based data in the PM context can imply errors in the described network of relations because of the occurrence of new environmental circumstances. On account of these changes, a resulting illustration of interdependencies can be inadequate to reality. Therefore, it is indispensable to improve these data with quantifying methods and consequently intersubjectivate them. So, there is a necessity of research in new mathematical applications with regard to measurement and especially to PM which is yet limited to the fundamental methodologies of sociology (survival analysis), psychology (various psychometric methods) and economics (econometrics) [11]. In such social economic systems with poor information, it is challenging to look for solutions in Statistics because of the system's dynamic characteristics. In this case of incomplete and fast-changing information, the application of GRA may be advisable [40].

3. Applying Grey systems in performance management

The GST first appeared in 1981 by Deng [41]. According to that, a Grey system (GS) has the structure of a black box, which contains a system of both known and unknown variables. The unknown represents a "black", totally incomplete information and the known a "white", absolutely complete information. Hence, a (Grey) incomplete information can be understood as an information that is partially known as well as to some extent unknown [42]. Inconsiderably, whether it is the message format, the coordination mechanism or just the behaviour within a system: As soon as a lack of information within this system is disseminated, it is referred to as a GS [36]. In practice, as already mentioned in the previous chapter, it is difficult to concretely obtain all information about an examined object [40]. Systems with a lack of information can be found everywhere: for example, the biological limitations of the human senses, the constraints of important economic conditions or the unavailability of technical resources. The GS as a system of incomplete information is also known as an "indeterminate system" of which the fundamental characteristics are small samples and/or interruptions of time series [42].

On the account of the small size of the samples problems within information systems with incomplete information cannot be solved with statistical methods [42]. With increasing sample size, the statistical power of a validation method grows [43]. Thus, sample sizes are preferable, in which the standard error is as low as possible. Various studies [44–46] consider large numbers of data points as necessary for the application of statistical support of time series as well as cross-sectional analysis in PM. For instance, according to McDonald and Ho [45], an organization needs to obtain quarterly data for a moderate time series analysis for almost six years in order to make a statement about possible causal relations by structural equation modelling. In social and economic systems, which are driven by the highest degree of dynamism and continuous changes, such problem solving demands for overextend the conditions of typical situations of business practice. Some variables in the system underlie a faster change of their environment conditions than the measurement lasts at all, so that the analytical results are irrelevant and therefore superfluous [41]. The resulting situation of incomplete information can be supported by GST [23].

The enormous volumes of data arising from subjective questionnaires about the performance indicators (k_i) need reduction. **Table 1** shows the result of such a decimation to those indicators which are most essentially interlinked with the financial performance generation. For this, benchmarking of the most representative indicators is crucial [47]. The expression x_{it} represents an opinion aggregated from the individual members of an expert group in period

t to performance indicator k_i. Here, GST disposes of a major advantage because of the ability to provide valid results already from a number of data points with $t \geq 4$. Thus, the GST is able to work with incomplete information in terms of decimating the indicators to the KPIs [36].

Period *t*						
Performance indicator (k_i)	Q_1	Q_2	Q_3	Q_4	…	Q_t
k_1	x_{11}	x_{12}	…	…	…	x_{1t}
k_2	…	x_{22}				…
k_3	…					…
k_4	…					…
…	…					…
k_i	x_{i1}	…	…	…	…	x_{it}

Table 1. Subjective questionnaire.

3.1. Buffer operator in Grey systems theory

The GST could be the way out for problems of incomplete and therefore inadequate data. The challenge for PM is especially the collection of performance relevant data often derived from the answers to subjective questionnaires within organizations. From time to time, this requires a certain number of subjective data as shown in **Table 1**. Nevertheless, in practice, it may occur that experts cannot answer their quarterly surveys (e.g., vacation, illness or simple absence). Therefore, the GST is providing a buffer operator, which makes it possible to complete missing information in fragmented queries, without this leading to informational distortion or loss. If two adjacent entries of a data sequence are described by $x(n-1)$ and $x(k)$, then, $x(k-1)$ represents an old information, and $x(k)$ operates as a part of a newer information. If there is a gap between entries within a data sequence, a lack of information because of the insufficient completion of an expert's questionnaire occurs (e.g., $X = (x(1), x(2)\ x(3), x(5))$. A new value $x(4)$ can be created as follows:

$$x^*(k) = \alpha \cdot x(k) + (1-\alpha) \cdot x(k-1), \alpha \in [0,1]. \quad (1)$$

The value of α represents the weighting of the informational content with regard to its currency. If $\alpha > 0.5$, the researcher attaches more importance to the newer information than to the older one and vice versa [23]. For simplification, no preference with respect to the timeliness of information should be assumed in the following, so that old and new information should be weighted equally ($\alpha = 0.5$).

In cases of a blank first entry $x(1)$ or a missing last entry $x(n)$ of a sequence X—for example, measured customer contentment—the gap cannot be filled by the method of adjacent neighbour generation, but rather by methods called stepwise ratio generator

$\sigma(k) = \frac{x(k)}{x(k-1)}; k = 2, 3, ..., n$ or the smooth ratio generator $\rho(k) = \frac{x(k)}{\sum_{i=1}^{k-1} x(i)}; k = 2, 3, ..., n$. If the first value is missing, the method operates with the adjacent values within the sequence right of the missing one: $x(1) = \frac{x(2)}{\sigma(3)}$ or $x(1) = \frac{x^2(2)}{x(3) - x(2)}$. If only the last sequence value shows an empty entry, the two previous sequence data help to create an adequate "substitute": $x(n) = x(n-1)\sigma(n-1)$ or $x(n) = x(n-1)(1 + \rho(n-1))$ [23].

3.2. Grey relation analysis

The challenge of GRA is to clarify which factors influence the PM system in a desirable extent, to strengthen and to focus those subsequently. In the past, this has been discussed in scientific articles and essays about system theory. However, this methodology still attends rare attention in the context of Performance Management [23, 48–50]. This model was chosen, as it tries to work as an ideal PM support with its consideration of both financially and nonfinancially dimensioned factors by analysing the system's factors that display sufficient influence on the top strategic financial ratio but appear as incomplete [51]. By means of the Performance Management as well as by the efficient and effective KPIs identified by the PM, the entire organization could be aligned to its strategy and vision [52]. Therefore, GRA attempts to discover the sequences of the KPIs by determining the geometrically most similar sequences to the top strategic financial performance ratio to uncover the system's most descriptive factors [23]. Therefore, an organization has to determine a reference sequence, which optimally represents the strategy of the organization and thus the behaviour of the entire system [53]. The strategy and hence the ultimate performance generation should be illustrated by the KPIs. Here, Paquette and Kida [27] showed in their study that it is important to reduce the extension of the strategy to a minimum. So, in order to reflect the strategy by a reference sequence, it is advisable to refer to a single factor and not to a variety of multiple sequences. Kasperskaya and Tayles [34] propose that both types of indicators (financial and nonfinancial) within a well-functioning PM system should be used, but, however, the financial measures dominate in practice. Kaplan and Norton [52] also consider that a financial measure should be attributed the most weight in a strategy-focused organization, so that it can monitor and control their operational and strategic budgeting. Thus, a financial measure should also be used as a reference sequence in the selection of the strategy-related KPIs in a PM System.

The GRA is a part of the GST mentioned earlier and is based on all of its assumptions and conditions [47]. In this context, a Grey relation proposes the valuation between two autonomous systems or two indicators within a system over a determined time series. It is precisely this point where the examination method GRA can be used. The elements are examined for homogeneous or heterogeneous temporal behaviour which means the development of the considered indicator in terms of time. If the elements display a very similar, homogeneous development concerning the time series, a high relational degree is assumed and vice versa. First, a reference sequence $X_0 = (x_0(1), x_0(2), ..., x_0(n))$ is defined. Afterwards it is possible, to compare the geometrical similarity of the reference sequence with another system's element and its sequence $X_i = (x_i(1), x_i(2), ..., x_i(n))$. If $\gamma(x_0(k), x_i(k))$ is the image of the real numbers at point k as well as $x_0(k)$ and $x_i(k)$ display x_0 and x_i at point k, and $\gamma(X_0, X_i)$ reflects all data points of every sequence (i = 1, 2, …, n) with k = 1, 2, …, m, then the Grey relation coefficient follows the formula [36, 53]:

$$\gamma(x_0(k), x_i(k)) = \frac{\min_i \min_k | x_0(k) - x_i(k) | + \xi \max_i \max_k | x_0(k) - x_i(k)}{| x_0(k) - x_i(k) | + \xi \max_i \max_k | x_0(k) - x_i(k) |}. \tag{2}$$

The value $\xi \in [0, 1]$ describes the differentiation coefficient that aids to adjust the various relation coefficients. Lin et al. [54] suggest for ξ a value of 0.5 to attain a stable and appropriate distinction.

Then, the Grey relational degree $\gamma(X_0, X_i)$ could be calculated as follows [36, 47, 53]:

$$\gamma(X_0, X_i) = \frac{1}{n} \sum_{k=1}^{n} \gamma(x_0(k), x_i(k)). \tag{3}$$

That leads to $0 < \gamma(X_0, X_i) < 1$, so that a value of 0 can be interpreted as a blank and a value of 1 suggests a complete and perfect relation of the two compared sequences [47].

Though, the Grey relational degree by Deng functions as a historical basis of the GRA in this case, it is not applied in the further process because of its dependence of the sequence order in the calculation $(\gamma(X_0, X_i) \neq \gamma(X_i, X_0))$, and therefore, its rank reversal problems. As a result, the more general Grey incidence analysis is focused which is based on the approaches of symmetry and thus protected against the problems of Deng's Grey relational degree. The relative and the absolute degree of incidence are to be attributed to the more general approach of the Grey incidence analysis. Nevertheless the Grey relational degree and the Grey incidence are used synonymously [53].

First, the absolute degree of incidence is considered. Assuming X_I is an economic factor of the regarded system and k represents the ordinal number of this factor. Then $X_i = (x_i(1), x_i(2), ..., x_i(n))$ represents the series of the index and thus the temporal behaviour of an economic factor. Equivalently, this could be transferred to the PM context, so that various cash flows of several months, confidence of the employees in management or contentment of customers with a corporate's products are constituted as performance indicators [24]. These sequences may take various forms. If the sequence of an indicator is given by $X_i = (x_i(1), x_i(2), ..., x_i(n))$, then $X_i - x_i(1)$, or rather:

$$\left(x_i(1)-x_i(1),\ x_i(2)-x_i(1),\ldots,\ x_i(n)-x_i(1)\right) \tag{4}$$

illustrates a fluctuating image and therefore a development of the indicator behaviour [23]. The area under the curve can therefore be quantified as follows:

$$s_i = \int_1^n \left(X_i - \left(x_i(1)\right)\right)\mathrm{dt}. \tag{5}$$

As a result, the sequences can occur by decreasing (A), increasing (B) and vibrating (C) temporal behaviour. To be able to compare sequences with each other, a zero-starting point operator is applied [23]:

$$X_i D = x_i(1)d, x_i(2)\mathrm{d}, \ldots, x_i(n)d \text{ and } x_i(k)d = x_i(k) - x_i(1) \text{ with } k = 1, 2, \ldots, n. \tag{6}$$

Consequently, the comparison of two sequences appears possible, so that also statements about the area beyond the curves can be made (**Figure 2**) [23].

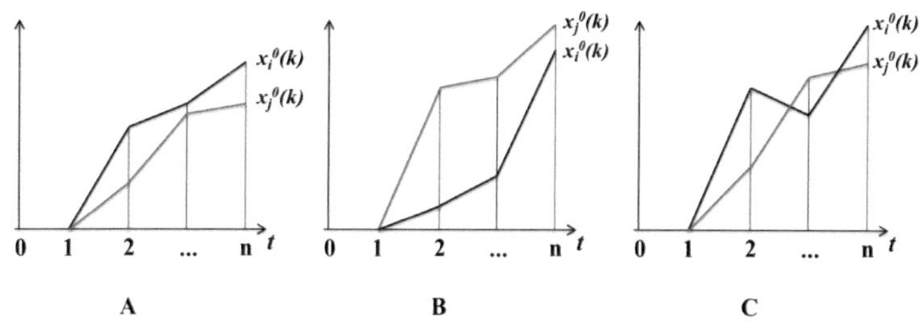

Figure 2. Relationship between two sequences. (A = x_i^0 is located above x_j^0; B = x_i^0 is located underneath x_j^0; C = x_i^0 and x_j^0 alternate positions).

Now, the area s_i between X_i^0 and abscissa can be calculated by the following equation [56]:

$$|s_i| = \left|\sum_{k=2}^{n-1} x_i^0(k) + \frac{1}{2}x_i^0(n)\right|. \tag{7}$$

Here, however, rather the area between the two curves, X_i^0 and X_j^0 is of interest, which can be described by the following equation [55]:

$$|s_{ij}| = |s_i - s_j| = \left| \sum_{k=2}^{n-1} \left(x_i^0(k) - x_j^0(k) \right) + \frac{1}{2}\left(x_i^0(n) - x_j^0(n) \right) \right| \tag{8}$$

Assuming the length of both sequences is the same (otherwise the sequences could be adjusted, as described in Subchapter 3.1), then the absolute degree of incidence of the sequences X_i and X_j can be determined by [23]:

$$\varepsilon_{ij} = \frac{1 + |s_i| + |s_j|}{1 + |s_i| + |s_j| + |s_i - s_j|}. \tag{9}$$

In the PM, the challenge is to associate both financial and nonfinancial measures [34]. However, in case of this endeavour, problems of differently scaled indicators can emerge. Likert scale estimations by experts of employee satisfaction, for example, can be set in relation. But this would be no normalization as demanded by the concept of the absolute degree of incidence [56]. On the contrary to this, the concept of the relative degree of incidence provides a quantitative description of the rate of change of two sequences to their initial values, thus enabling a sufficient normalization. The closer these rates of changes of the two sequences are, the greater is the relative degree of incidence r_{ij} between them. Assuming that X_i and X_j are two sequences of an equal length with initial values that are different to zero, then there is no connection or linkage between the absolute and the relative degree of incidence, so that the absolute degree ε_{ij} can be relatively large, whereas its relative counterpart r_{ij} can be extremely small and vice versa [23]. Using the relative degree of incidence, an equal length of the sequences is assumed, this means an identical number of data points for the two sequences X_0 and X_i, where $X_0 = (x_0(1), x_0(2), ..., x_0(k))$ constitutes the reference sequence. Afterwards, to be able to compare the possibly differently scaled indicators and so their sequences, the values of each sequence are divided by their initial value:

$$X_0' = \left(x_0'(1), x_0'(2), \ldots, x_0'(k) \right) = \frac{x_0(1)}{x_0(1)}, \frac{x_0(2)}{x_0(1)}, \ldots, \frac{x_0(k)}{x_0(1)}, \tag{10}$$

$$X_i' = \left(x_i'(1), x_i'(2), \ldots, x_i'(k) \right) = \frac{x_i(1)}{x_i(1)}, \frac{x_i(2)}{x_i(1)}, \ldots, \frac{x_i(k)}{x_i(1)}. \tag{11}$$

Subsequently, the zero-starting point is determined analogously to Eq. (5), so there is the possibility to calculate the areas $|s_i|$ and $|s_{ij}|$ as well as the relative degree of incidence r_{ij} [57]:

$$|s_i'| = \left| \sum_{k=2}^{n-1} x_i'^0(k) + \frac{1}{2} x_i'^0(n) \right|, \tag{12}$$

$$|s_{ij}'| = \left| s_i' - s_j' \right| = \left| \sum_{k=2}^{n-1} \left(x_i'^0(k) - x_j'^0(k) \right) + \frac{1}{2} \left(x_i'^0(n) - x_j'^0(n) \right) \right|, \tag{13}$$

$$r_{ij} = \frac{1 + |s_i'| + |s_j'|}{1 + |s_i'| + |s_j'| + |s_i' - s_j'|}. \tag{14}$$

Using these formulas, it is possible to calculate the respective relative degree of incidence between the variety of performance indicators and the reference sequence, to disclose for example the ten most "important" sequences/indicators for the reference sequence, the KPIs. To get an overview of the dependencies within those 10 KPIs, also the relative degrees of incidence between the KPIs can be calculated so that an interdependency network emerges [55]. Since there is only the possibility of building a network of interdependencies between the KPIs by GRA, the cause-and-effect-relationships lack a detailed explanation. This network, however, is likely to be understood as a construct of the holistic organizational strategy which is determined by "highly correlated" KPIs. If then the strategy changes or rather is adjusted to altered circumstances, the indicators act to the same extent, so that their cause-and-effect relationships are inconsiderable [58]. Nevertheless, the KPIs in their combination must be selected providing sufficiently the strategy and therefore its means and relations.

4. Example of application

The following example of a PM relevant application shall illustrate the possibility of simplifying the indicator selection in the PM with GRA in case of poor data situations. Therefore, the estimations of 50 performance indicators, the possible KPIs, by five organizational experts over four quarters serve as initial data for the example. For the reference sequence, to reflect the corporate strategy as simple as possible, the cash flows over the four quarters are used. The 50 performance indicators show a pre-selected pool of indicators elicited, for example by interviews [10]. They can range from employee satisfaction over customer contentment to process quality, for instance. Then, the experts are encouraged to estimate the respective extent of the indicator k_{it} in the considered period with regard to the Saaty scale (with 1 = very weak extent to 9 = very strong extent) [59]. After the other four experts have analogously estimated, the respective indicators in each period, an aggregated group matrix is created by the mean value of the experts' estimations (**Table 2**). The corresponding cash flows of the considered periods should fictitiously serve as a compliant financial target indicator of the corporation and thus as the reference sequence of the application example.

According to the equal length of all sequences, the values of **Table 3** can be normalized in a certain way by Eq. (10) in order to make the differently scaled sequences comparable (**Table 3**).

The indicators 1–50 do not require to consist of subjective data. For example, customer satisfaction, as a performance indicator, could be represented by an objective measure such as the amount of product returns, if existing. Subsequently, the sequences of **Table 3** need to be moved to an initial value of zero with the zero-starting point operator of Eq. (6) (**Table 4**).

Aggregated experts' estimations	**Period** ***t***	Q_1	Q_2	Q_3	Q_4
Reference sequence j: cash flow		1,000,000	1,500,000	1,750,000	1,250,000
Performance indicator (k_i)					
k_1		4.0000	4.4000	2.6000	4.6000
k_2		5.4000	6.6000	4.8000	2.8000
k_3		5.0000	4.0000	3.4000	5.2000
k_4		7.0000	5.4000	4.8000	3.0000
…		…	…	…	…
k_{50}		5.2000	5.6000	5.4000	4.4000

Table 2. Aggregated experts' estimations.

Normalized aggregated estimations	**Period** ***t***	Q_1	Q_2	Q_3	Q_4
Reference sequence j: cash flow		1.0000	1.5000	1.7500	1.2500
Performance indicator (k_i)					
k_1		1.0000	1.1000	0.6500	1.1500
k_2		1.0000	1.2222	0.8889	0.5185
k_3		1.0000	0.8000	0.6800	1.0400
k_4		1.0000	0.7714	0.6857	0.4286
…		…	…	…	…
k_{50}		1.0000	1.0769	1.0385	0.8462

Table 3. Normalized aggregated estimations.

Then, it is possible to calculate the area between the abscissa and the respective sequence $|s_i'|$ by Eq. (12). The geometrical nearness between a considered sequence and the cash flow

http://dx.doi.org/10.5772/65286

reference sequence $|s'_{ij}|$ can be determined by Eq. (13) and consequently also the relative degree of incidence r_{ij} with the help of Eq. (14).

Thus, it is possible to provide a ranking of the geometrically most similar sequences with regard to the cash flow reference sequence (**Table 5**). In this example, the number of KPIs is limited to a count of ten as proposed by Markóczy and Goldberg as the optimal number to work with in PM [60].

GRA not only provides a ranking of the most important indicators of complex systems, it also offers the possibility to reveal the dependencies between the considered indicators by a network map. For this purpose, the relative degrees of incidence between the ten KPIs are determined by Eq. (13) (**Table 6**).

Images with zero-starting point	**Period *t***	Q_1	Q_2	Q_3	Q_4	$\lvert s_i \rvert$	r_{ij}
Reference sequence j: cash flow		0.0000	0.5000	0.7500	0.2500	1.6250	1.0000
Performance indicator (k_i)							
k_1		0.0000	0.1000	−0.3500	0.1500	0.0250	0.3846
k_2		0.0000	0.2222	−0.1111	−0.4815	0.0611	0.6207
k_3		0.0000	−0.2000	−0.3200	0.0400	0.4600	0.6969
k_4		0.0000	−0.2286	−0.3143	−0.5714	1.4000	0.5590
…		…	…	…	…	…	
k_{50}		0.0000	0.0769	0.0385	−0.1538	0.1154	0.4056

Table 4. Images with zero-starting point.

Key performance indicator	**Relative degree of incidence (r_{ij})**	**Ranking**
k_{15}	0.9219	1
k_{10}	0.9167	2
k_{14}	0.9129	3
k_{13}	0.8958	4
k_{23}	0.8857	5
k_{21}	0.8847	6
k_{31}	0.8552	7
k_{43}	0.7780	8
k_{24}	0.7719	9
k_{12}	0.7572	10

Table 5. Relative degrees of incidence of the performance indicators and their ranking.

KPI	k_{15}	k_{10}	k_{14}	k_{13}	k_{23}	k_{21}	k_{31}	k_{43}	k_{24}	k_{12}
k_{15}	1.0000	0.4995	0.5629	0.9219	0.8400	0.9422	0.7285	0.9904	0.5635	0.5948
k_{10}		1.0000	0.8161	0.4793	0.5520	0.5153	0.6138	0.5020	0.8148	0.7572
k_{14}			1.0000	0.5373	0.6305	0.5830	0.7123	0.5660	0.9980	0.9129
k_{13}				1.0000	0.7842	0.8725	0.6862	0.9137	0.5378	0.5663
k_{23}					1.0000	0.8857	0.8458	0.8469	0.6312	0.6708
k_{21}						1.0000	0.7626	0.9508	0.5837	0.6174
k_{31}							1.0000	0.7337	0.7133	0.7642
k_{43}								1.0000	0.5666	0.5983
k_{24}									1.0000	0.9145
k_{12}										1.0000

Table 6. Network of KPI dependencies.

Table 6 shows the relative degrees of incidence between the KPIs, which can be interpreted as reciprocal as these degrees can be understood as a kind of "Grey Correlation" [42]. However, similar to the DEMATEL approach, it is important to limit the dependencies to the really "essential" and "significant" ones. Therefore, the shaded fields are not considered subsequently so that only those dependencies which exceed the threshold, the average of the matrix (mean value = 0.74161862), should remain for further analytical procedure.

5. Results and conclusion

The GST shows considerable advantages, particularly in a complex system as the PM. At the present time, it is indispensable to involve the dynamic environment in management control. For this purpose, it is necessary to continuously focus the corporate strategy and objectives in order to create a long-term financial success. The problems that especially occur as a consequence of incomplete information and small sample sizes can be a huge hurdle. The PM requires a permanent update which cannot be enabled by mere application of the existing statistical methods. The PM represents a highly dynamical system with ever-changing environmental conditions. This prohibits an appropriate data measurement with analysis by common statistical methods. Data alter before statistic samples can provide any analytic results. Therefore, it is important to seek methods with minimum data size demands. According to that, the GST with its applications can be useful with its low requirements in sample sizes. Specifically, GRA offers important advantages for the selection of KPIs in poor data situations with the additional possibility of a visual representation of the revealed KPIs within a network of interdependencies.

In conclusion, GRA provides the feasibility to support the performance generation process and to assist PM as a tool-selecting performance indicators in case of incomplete information with small sample sizes. Besides, GRA is able to visualize the performance generation in a map that facilitates steering and control of the organization in the framework of Performance Management [35]. The ability to include financial and non-financial measures provides further advantages for GRA. So, it definitely appears suitable as an OR tool for management control, in particular in PM.

GRA as one of the submethods of the GST will help to improve the informational bases of PM by its possibilities of flexible usage. Therefore, GRA should serve as a feedback as well as a feedforward-oriented PM support. Initially, it provides intersubjectivated data for the performance management, which then disposes of improved informational bases for counteraction measures. After structural breaks of the system, in which PM is implemented, GRA is supposed to inform about such defects and should operate as a feedforward-oriented support for deriving, validating and implementing a new causal model.

The rising number of OR-publications on GST issues demonstrates the enhancing importance of this theory for the analysis of complex systems. However, there are only a small number of articles in the PM literature referring to GST [49]. GST with its wide range of applications is nevertheless an appropriate OR method to support PM. Because of its relevance specifically in poor data situations with incomplete information, PM literature should increasingly focus GST as an important support instrument.

Author details

Thorben Hustedt, Wolfgang Ossadnik* and Fabian Burrey

*Address all correspondence to: wolfgang.ossadnik@uni-osnabrueck.de

Department of Management Science/Management Accounting and Control, University of Osnabrück, Osnabrück, Germany

References

[1] Rigby D, Bilodeau B. Management tools and trends 2013. Bain and Company. 2013;1:1–12.

[2] Lebas M J. Performance measurement and performance management. International Journal of Production Economics. 1995;41(1–3):23–35. DOI: 10.1016/0925-5273(95)00081-X.

[3] Ghalayini A M, Noble J S, Crowe T J. An integrated dynamic performance measurement system for improving manufacturing competitiveness. International Journal of Production Economics. 1997;48(3):207–225. DOI: 10.1016/S0925-5273(96)00093-X.

[4] Ossadnik W. Performance management of internationally acting firms as a challenge for management accounting and control and stimulant of their further development [performance management international agierender Unternehmen als Herausforder-

ung für die Controllinglehre und Stimulanz für deren Weiterentwicklung] (to be published). In: Funk W, Rossmanith J, editors. International Accounting and International Controlling [Internationale Rechnungslegung und Internationales Controlling]. 3rd ed. Wiesbaden: Gabler; 2017. p. 1–31.

[5] Ghalayini A M, Noble J S. The changing basis of performance measurement. International Journal of Operations and Production Management. 1996;16(8):63–80. DOI: 10.1108/01443579610125787.

[6] Marr B, Schiuma G, Neely A. Intellectual capital—defining key performance indicators for organizational knowledge assets. Business Process Management Journal. 2004;10(5):551–569. DOI: 10.1108/14637150410559225.

[7] Neely A, Gregory M, Platts K. Performance measurement system design: a literature review and research agenda. International Journal of Operations and Production Management. 1995;15(4):80–116. DOI: 10.1108/01443579510083622.

[8] Nørreklit H. The balance on the balanced scorecard—a critical analysis of some of its assumptions. Management Accounting Research. 2000;11(1):65–88. DOI: 10.1006/mare.1999.0121.

[9] Buytendijk F, Hatch T, Micheli P. Scenario-based strategy maps. Business Horizons. 2010;53(4):335–347. DOI: 10.1016/j.bushor.2010.02.002.

[10] Abernethy M A, Horne M, Lillis A M, Malina M A, Selto F H. A multi-method approach to building causal performance maps from expert knowledge. Management Accounting Research. 2005;16(2):135–155. DOI: 10.1016/j.mar.2005.03.003.

[11] Richard P J, Devinney T M, Yip G S, Johnson G. Measuring organizational performance: towards methodological best practice. Journal of Management. 2009;35(3):718–804. DOI: 10.1177/0149206308330560.

[12] Chytas P, Glykas M, Valiris G. A proactive balanced scorecard. International Journal of Information Management. 2011;31(5):460–468. DOI: 10.1016/j.ijinfomgt.2010.12.007.

[13] Ambrosini V, Bowman C. Tacit knowledge: some suggestions for operationalization. Journal of Management Studies. 2001;38(6):811–829. DOI: 10.1111/1467-6486.00260.

[14] Nonaka I. A dynamic theory of organizational knowledge creation. Organization Science. 1994;5(1):14–37. DOI: 10.1287/orsc.5.1.14.

[15] Bretzke W R. The Problem Reference of Decision Models [Der Problembezug von Entscheidungsmodellen]. 1st ed. Tübingen: Mohr Siebeck; 1980. p. 280.

[16] Bourne M, Mills J, Wilcox M, Neely A, Platts K. Designing, implementing and updating performance measurement systems. International Journal of Operations and Production Management. 2000;20(7):754–771. DOI: 10.1108/01443570010330739.

[17] Simons R, Dávila A. How high is your return on management?. Harvard Business Review. 1998;76(1):70–80.

[18] Heneman R L, Ledford Jr. G E, Gresham M T. The changing nature of work and its effects on compensation design and delivery. In: Heneman R L, editor. Strategic Reward Management: Design, Implementation, and Evaluation. 1st ed. Greenwich, CT: Information Age Publishing; 2002. p. 35–74.

[19] Schick A G, Gordon L A. Information overload: a temporal approach. Accounting, Organizations and Society. 1990;15(3):199–220. DOI: 10.1016/0361-3682(90)90005-F.

[20] Ittner C D, Larcker D F, Randall T. Performance implications of strategic performance measurement in financial firms. Accounting, Organizations and Society. 2003;28(7-8): 715–741. DOI: 1016/S0361-3682(03)00033-3.

[21] Mosakowski E. Strategy making under causal ambiguity: conceptual issues and empirical evidence. Organization Science. 1997;8(4):414–442. DOI: 10.1287/orsc.8.4.414

[22] Salmeron J L. Modelling grey uncertainty with Fuzzy Grey Cognitive Maps. Expert Systems with Applications. 2010;37(12):7581-7588. DOI: 10.1016/j.eswa.2010.04.085.

[23] Liu S, Lin Y. Grey Systems—Theory and Applications. 1st ed. Berlin, Heidelberg: Springer; 2010. p. 379.

[24] Schreyögg G, Steinmann H. Strategic control: a new perspective. Academy of Management Review. 1987;12(1):91–103. DOI: 10.5465/AMR.1987.4306487

[25] Daft R L, Lengel R H. Information richness: anew approach to managerial behavior and organization design. In: Staw B M, Cummings L L, editors. Research in Organizational Behavior. 6th ed. Greenwich, CT: JAI Press; 1984. p. 191–233.

[26] Zimmermann H-J. An application-oriented view of modeling uncertainty. European Journal of Operational Research. 2000;122(2):190–198. DOI: 10.1016/S0377-2217(99)00228-3.

[27] Paquette L, Kida T. The effect of decision strategy and task complexity on decision performance. Organizational Behavior and Human Decision Processes. 1988;41(1):128–142. DOI: 10.1016/0749-5978(88)90051-9.

[28] Akkermans H, van Helden K. Vicious and virtuous cycles in ERP: implementation: a case study of interrelations between critical success factors. European Journal of Information Systems. 2002;11(1):35–46. DOI: 10.1057/palgrave.ejis.3000418.

[29] Kruis A-M, Speklé R F, Widener S K. The levers of control framework: an exploratory analysis of balance. Management Accounting Research. Forthcoming. DOI: 10.1016/j.mar.2015.12.002.

[30] Eden C, Ackermann F. The analysis of cause maps. Journal of Management Studies. 1992;29(3):309–324. DOI: 10.1111/j.1467-6486.1992.tb00667.x.

[31] Lebas M, Euske K. A conceptual and operational delineation of performance. In: Neely A, editor. Business Performance Measurement. Theory and Practice. 1st ed. Cambridge: Cambridge University Press; 2002. p. 65–79.

[32] Axelrod R. Structure of Decision: The Cognitive Maps of Political Elites. 1st ed. Princeton, NJ: Princeton University Press; 1976. p. 422.

[33] Fiol C M, Huff A S. Maps for Managers: Where are we? Where do we go from here?. Journal of Management Studies. 1992;29(3):267–285. DOI: 10.1111/j.1467-6486.1992.tb00665.x.

[34] Kasperskaya Y, Tayles M. The role of causal links in performance measurement models. Managerial Auditing Journal. 2013;28(5):426–443. DOI: 10.1108/02686901311327209.

[35] Bititci U, Cocca P, Ates A. Impact of visual performance management systems on the performance management practices of organisations. International Journal of Production Research. 2016;54(6):1571–1593. DOI: 10.1080/00207543.2015.1005770.

[36] Deng J. Introduction to grey system theory. The Journal of Grey System. 1989;1(1):1–24.

[37] Nørreklit H, Mitchell F. The balanced scorecard. In: Hopper T, Northcott D, Scapens R, editors. Issues in Management Accounting. 3rd ed. London: Prentice Hall; 2007. p. 175–198.

[38] Speckbacher G, Bischof J, Pfeiffer T. A descriptive analysis on the implementation of balanced scorecards in German-speaking countries. Management Accounting Research. 2003;14(4):361–388. DOI: 10.1016/j.mar.2003.10.001.

[39] Kelly K. Accuracy of relative weights on multiple leading performance measures: effects on managerial performance and knowledge. Contemporary Accounting Research. 2010;27(2):577–608. DOI: 10.1111/j.1911-3846.2010.01017.x.

[40] Chen M-Y, Li Z, Zhou L, Xiong H, An X. SCGM-model and grey control of "poor" information systems. Kybernetes. 2004;33(2):231–237. DOI: 10.1108/03684920410514157.

[41] Deng J. Control problems of grey systems. Systems and Control Letters. 1982;1(5):288–294.

[42] Lin Y, Chen M-Y, Liu S. Theory of grey systems: capturing uncertainties of grey information. Kybernetes. 2004;33(2):196–218. DOI: 10.1108/03684920410514139.

[43] Gujarati D N, Porter D C. Basic Econometrics. 5th ed. Singapore: McGraw-Hill Education Ltd.; 2009. p. 922.

[44] Kelley K, Maxwell S E. Sample size for multiple regression: obtaining regression coefficients that are accurate, not simply significant. Psychological Methods. 2003;8(3): 305–321. DOI: 10.1037/1082-989X.8.3.305

[45] McDonald R P, Ho M-H R. Principles and practice in reporting structural equation analyses. Psychological Methods. 2002;7(1):64–82. DOI: 10.1037//1082-989X.7.1.64.

[46] Anderson J C, Gerbing D W. Structural equation modeling in practice: a review and recommended two-step approach. Psychological Bulletin. 1988;103(3):411–423. DOI: 10.1037/0033-2909.103.3.411.

[47] Ho C-T, Wu Y-S. Benchmarking performance indicators for banks. Benchmarking: An International Journal. 2006;13(1/2):147–159. DOI: 10.1108/14635770610644646.

[48] Yin M-S. Fifteen years of grey system theory research: a historical review and bibliometric analysis. Expert Systems with Applications. 2013;40(7):2767–2775. DOI: 10.1016/j.eswa.2012.11.002.

[49] Delcea C. Grey systems theory in economics—bibliometric analysis and applications' overview. Grey Systems: Theory and Application. 2015;5(2):244–262. DOI: 10.1108/GS-03-2015-0005.

[50] Kennerly M, Neely A. Measuring performance in a changing business environment. International Journal of Operations and Production Management. 2003;23(2):213–229. DOI: 10.1108/01443570310458465.

[51] Kaplan S E, Petersen M J, Samuels J A. An examination of the effect of positive and negative performance on the relative weighting of strategically and non-strategically linked balanced scorecard measures. Behavioral Research in Accounting. 2012;24(2): 133–151. DOI: 10.2308/bria-50114.

[52] Kaplan R S, Norton D P. The Strategy-Focused Organization. 1st ed. Cambridge, MA: Harvard Business Review Press; 2000. p. 416.

[53] Liu S, Yang Y, Cao Y, Xie N. A summary on the research of GRA models. Grey Systems: Theory and Application. 2013;3(1):7–15. DOI: 10.1108/20439371311293651.

[54] Lin S J, Lu I J, Lewis C. Grey relation performance correlations among economics, energy use and carbon dioxide emission in Taiwan. Energy Policy. 2007;35(3):1948–1955. DOI: 10.1016/j.enpol.2006.06.012.

[55] Tung C-T, Lee Y-J. The innovative performance evaluation model of grey factor analysis: a case study of listed biotechnology corporations in Taiwan. Expert Systems with Applications. 2010;37(12):7844–7851. DOI: 10.1016/j.eswa.2010.04.053.

[56] Lin C-T, Tsai H-Y. Hierarchical clustering analysis based on grey relation grade. Information and Management Sciences. 2005;16(1):95–105.

[57] Scarlat E, Delcea C, Maracine V. Genetic—fuzzy grey algorithms: a hybrid model for establishing companies' failure reasons. In: IEEE, editor. 2010 IEEE International Conference on Systems Man and Cybernetics (SMC); 10–13 Oct. 2010; Istanbul. Istanbul: IEEE; 2010. p. 955–962.

[58] Albers S, Hildebrandt L. Methodological problems in research of success factors: measurement errors, formative versus reflective indicators and the choice of the structural equation model [Methodische probleme bei der Erfolgsfaktorenforschung - Messfehler, formative versus reflektive Indikatoren und die Wahl des Struktur-

gleichungs-Modells]. Schmalenbachs Zeitschrift für betriebswirtschaftliche Forschung. 2006;58(1):2–33. DOI: 10.1007/BF03371642

[59] Saaty T L. Decision Making—The analytical hierarchy and network processes (AHP/ANP). Journal of Systems Science and Systems Engineering. 2004;13(1):1–35. DOI: 10.1007/s11518-006-0151-5.

[60] Markóczy L, Goldberg J. A method for eliciting and comparing causal maps. Journal of Management. 1995;21(2):305–333. DOI: 10.1177/014920639502100207.

Chapter 3

Application of Lean Methodologies in a Neurosurgery High Dependency Unit

Ricardo Balau Esteves, Susana Garrido Azevedo and Francisco Proença Brójo

Additional information is available at the end of the chapter

http://dx.doi.org/10.5772/64715

Abstract

This study aims to apply Lean methodologies at a neurosurgery high dependency unit (NHDU) for increasing safety and quality on the care delivered to acute neuropatients and to reduce time, steps, and distance travelled by nurses accessing life support equipment (LSE). The methodology used in this study is an action research, supported by a longitudinal mixed method approach with a one-group within-subjects pretest-posttest experimental type design. Resulting in a high waste of time, steps, and distance travelled to reach them. After the application of Lean methodologies, distance, steps, and time travelled by Nurses were quite improved. Lean methodologies applied in NHDU contributed to improve the organization, availability, and accessibility of LSE by putting them at the point-of-use. Quality and safety of patient care were also improved by allowing almost immediate life support interventions. Resistance to change was the major limitation. The Lean philosophy empowers health facility managers with tools and methodologies that help them create health gains, implement a culture of continuous improvement of care and working environment, identify and eliminate barriers, and waste that limits the work of staff in providing quality services and saving lives. This chapter highlights the responsibility of health facility managers to properly organize health units to cope with emergency situations, by allowing immediate, efficient, and effective intervention of staff.

Keywords: lean methodologies, critical care nursing, management, work simplification, action research

1. Introduction

Imagine that in a ward or in an acute care unit, a patient develops a sudden and severe laryngeal edema and stops breathing for obstruction of the respiratory tract. Nurse and medical staff starts-up advanced life support (ALS) maneuvers. The primary and emerging interventions are to permeabilize the airway accessing the trachea through an endotracheal tube. Not being possible to access it by the usual routes, the only solution is to perform a tracheotomy or cricothyrotomy, using a tracheotomy surgical tray (TST) or an emergency cricothyrotomy kit (ECK). However, the health team most of the times do not know its existence, location, or has difficulties in accessing it in due time, what could have as consequence the loss of a life.

Portuguese Directorate-General of Health (DGS) points out on the Circular Normative no. 15/DQS/DQCO of 22/06/2010 [1] that "patients who are admitted in hospitals believe that they are being admitted to a safe environment. They feel confident that if their clinical condition gets worse, they are in the best place for a prompt and effective intervention. However, there is some evidence that this does not always happen" (p. 1). This Circular also states, "ALL inpatient areas should have easy and immediate access to equipment, supplies and emergency drugs. They should be organized and stored in a standardized way… throughout the health unit" (p.6). However, compliance with these recommendations depends, above all, on political and management decisions of legislators, regulators, managers, and industry providers, and it is at the health institutions jurisdiction to "adequate resources and create the structures that leads to quality professional practice" [2]. For the Portuguese Republic Government (Governo da República Portuguesa—GRP) is "fundamental, that the available resources are better used, avoiding waste, that is, improving management, transparency, and accountability for the use of money from the citizens" [3]. Corvi [4] draws attention to the "waste epidemic in health care," as acknowledge by the GRP, and so a great opportunity to improve, being fundamental a spirit of continuous learning as part of "implementing a lean management system" [4].

The Intensive Care Society [5] recommends "all critical care areas should have their own, appropriately stocked and checked difficult airway trolley to deal with airway and tracheostomy emergencies" (p.11). The absence or inaccessibility to this kind of equipment can lead to adverse events with huge impact on the safety and lives of patients, mainly the critical ones. To the matter of the impact of layout configurations in hospital environment, Soriano-Meier et al. [6] points out that "inadequate facility layout negatively affects the performance of the service staff, the quality of care provision and the service temporally over time" (p. 255).

At a particular neurosurgery high dependency unit (NHDU) of a central hospital in Lisbon, problems related to design, layout, architectural barriers, accessibility to life support equipment (LSE) and wastes of time, handling, and transport were identified. Those that may infer greater impact on patients and on the provision of care taken by nurses are as follows: (a) accessibility to LSE, (b) difficulty/inexperience in the use of the resuscitation trolley (RT), and (c) lack of knowledge of ECK and TST existence and location. This chapter summarizes the action research study embarked with the purpose of testing the application of Lean methodologies at NHDU for a quality and safety provision of care to acute neuropatients and to reduce at least by half, time, steps, and distance travelled by nurses' accessing LSE. Gemba walk, value stream mapping, spaghetti diagrams, 5S, and JIT (just-in-time) were the main Lean methodologies used.

Gemba is the Japanese term used for Shop floor, where products are produced, or where the services are provided [7, 8]. To start an improvement project, it is critical to analyze the current and real situation of an organization or its workplace. Therefore, the gemba should analyze processes, time of setups, physical layout and surroundings with an open mind, to detect where, how, and why clients and staff experience problems [7–9].

The value stream mapping (VSM) is one of the main lean methodologies that engages waste elimination in any organization [10]. VSM will help to identify and analyze, for example, problems experienced by stakeholders, errors in medicine, flow of processes and work, financial analysis, among others. VSM allows checking (visual and graphically) the current state of a particular procedure, its productive time (value-added), and the non-productive time (non-value-added) [11].

Jackson [12] argues that 5S is the foundation of the Toyota production system (TPS). What this methodology tries to ensure is an orderly organized workspace for an efficient and safe work environment [10], increased productivity, fewer errors, and less waste [8]. 5S represents the five levels of this methodology starting with the letter "S". In Japanese vocabulary: SEIRI (sort), SEITON (set in order), SEISO (shine), SEIKETSU (standardize), and SHITSUKE (sustain). Smart [13] summarizes this methodology with the expression "a place for everything and everything in its place" (p. 62).

JIT is a production process that targets the optimization of the process as a whole in a continuous flow improvement and tries to answer to the organization or service needs. Briefly, it means producing no sooner, no later, neither more or less, only and just the necessary [8, 10].

Another Lean methodology is the spaghetti diagram. This diagram consists on a graphic reproduction of the architectural floor plan of a structure, where you draw lines from one space to another, representing the path taken by employees, customers, and objects along a particular process (round trip) [12, 14]. It allows documenting and visualizing the physical flow, in order to identify waste motion or transportation, architectural barriers, and improvement opportunities to expedite process flow [15].

This chapter is organized as follows: Following the introduction, a literature review on Lean philosophy is performed, then the methodology used in the research, and the results of the action research are described. Finally, some discussion and conclusions are drawn.

2. Lean philosophy

It was through Krafcik [16] that the Lean term was released thus referring to the TPS as a lean production system: A system that uses less resources compared to the mass production systems. Less effort, less capital investment, less space, and less time [17]. The Lean philosophy is essentially focused on waste reduction as a means to increase actual value-added, in order to fulfill customer needs and maintain profitability [17]. The fundamental focuses of Lean are respect for people, teamwork, waste elimination, continuous improvement, value, quality, and safety [8, 16–18]. Several authors have highlighted this and other key principles of Lean philosophy, such as follows: (i) customer relationship [19]; (ii) total quality management (TQM) [20]; (iii) JIT [21, 22]; (iv) pull production/flow [19, 20]; (v) supplier relationships/long-term business relationship [21]; (vi) mistake-proofing [23]; (vii) total productive maintenance (TPM) [22]; and (viii) physical layout [6]. At the operational level, the Lean paradigm is implemented using a number of techniques such as kanban, 5S, visual control, takt-time, poke-yoke, and single minute exchange of die (SMED) [24].

To Imai [8], the importance of applying a Lean philosophy in an organization has at least three components: (1) any activity or process that does not add value is waste, independently of being practiced by people or machines; (2) the reduction or elimination of waste may be the most cost-effective way for improving productivity and reduce operating costs instead of increasing investment in the hope of adding value. Moreover, investing in new equipment is expensive while eliminating waste, in most cases, has no costs. (3) Standardization of processes ensures quality and error prevention. Womack et al. [17] documented the benefits of Lean philosophy compared to the mass production model, arguing that this philosophy would succeed, not only in the automotive industry or aviation, but also in all activities from distribution, retail, and healthcare. Not being the solution to all the problems that health services faces today, the Lean philosophy can bring significant benefits to this sector and in a range of hospital areas [25], contributing to develop the continuous improvement into the organizational culture and improving quality of care, efficiency and effectiveness, while reducing costs, errors, and waste.

In the Portuguese healthcare sector, the implementation of Lean philosophy has been focused on some specific areas such as quality [26, 27] logistics, supply and storage [28–30], agility and continuous process improvement [31–33], workplace reorganization [34], and reducing waiting times [35, 36]. Particularly in services such as community health centers [37], operating room, imaging, ophthalmology, outpatient, ward, pharmacy, and warehouse. Other studies focused on conducting systematic reviews [38, 39]. There is thus a research gap in applying the Lean philosophy to inhospital medical emergency, especially in inpatient critical care services.

3. Methodology

The methodology used in this study was an action research, supported by a longitudinal mixed method approach with a one-group within-subjects pretest-posttest experimental type design.

Lewin [40] suggests the existence of a cycle in action research. It begins with the diagnosis and identification of the problem(s) with all participants in a democratic way and then follows the proposal, planning interventions, and actions of change. Subsequently, the impact of the changes is monitored, the data collected, analyzed, interpreted, and finally results are reported.

http://dx.doi.org/10.5772/64715

This is a flexible research methodology that integrates an exploratory action in order to investigate and support the implementation of changes according to the diagnosis raised [41]. Action research claims that the researcher participates in the change process since the changes suggested are implemented by himself, that is, he "take action to improve the practice and study … the effects of the action taken" [42]. Yin [43] considers this methodology as a variant of qualitative research that emphasizes the researcher action role and his active collaboration with the research participants.

3.1. Research design

The research was performed at a level 2 patient care four-bedded NHDU. This unit shares human resources, equipment, and materials with the 44-bedded standard care neurosurgery and neurotraumatology wards. NHDU is a healthcare facility specialized in the care of neuropatients undergoing neurological, hemodynamic, and respiratory instability with the eventual need of non-invasive or invasive ventilatory support by tracheotomy. These patients require critical care nursing and permanent vigilance that, although not requiring intensive care, may potential and quickly evolve to a severe status and thus the need of an immediate intervention. Nurse:patient ratio is 1:4. Located in one of an 802-bedded triple hospital centre at the metropolitan area of Lisbon (Portugal), this centre serves about a million people population. Data available from 2013 institutional performance reports show a surgical movement of 1423 neurosurgeries and a bed occupancy rate of 87.7% and 91.4% at the neurotrauma and neurosurgery wards, respectively.

The research was authorized by the NHDU Medical Director, the NHDU Chief Nurse, and the Ethics for Health Committee of the hospital centre. The unit of analysis is the NHDU with the corresponding nurse team. A convenience sampling was used attending to nurses' availability during the period that took place the visit of the researcher. The two nurses of the management team (chief and coordinator) were excluded from this sample since the purpose was to simulate the performance of the direct care nurses. Thus, from a population of 20 nurses, a sample of 12 nurses (60%) was selected. This is a longitudinal research in which data were collected from two points in time, which allowed studying the changes that have occurred during the period in which it was conducted (November 2014 to January 2015).

The research design follows several phases. The main three phases were (1) pre-intervention, (2) intervention, and (3) post-intervention, in which a simultaneous mixed method approach (qualitative and quantitative) was applied. The pre-intervention phase was further divided into three sub-phases: (i) diagnostic assessment (qualitative approach), (ii) simulation (quantitative approach), and (iii) proposal of changes (qualitative approach). The intervention phase consisted on the application of 5S and JIT lean methodologies. The post-intervention phase was divided into two qualitative approaches: (i) simulation and (ii) unstructured interview.

The pre-intervention diagnostic assessment sub-phase involved the following activities: (a) direct observation of the physical space performed by the participant researcher (PR) which focused mainly on the layout of the NHDU and the location of existing materials and equipment. To support the gemba walk, pictures and paper record with graphical representation of the service plan were used to complement the visual management and spaghetti diagram. The transition to digital record was made using Microsoft® Office® 2013 software. (b) Personal unstructured interviews performed by the PR to the nurse team, and questionnaires to identify the difficulties and constraints of nurses in their professional daily routines, especially in emergency situations. The questionnaires were anonymous and blind in order to guarantee their confidentiality. The participants returned them in a sealed envelope deposited at a container left in the nursing room. The analysis of questionnaires and interviews was performed using qualitative content analysis, and it was organized according to the research variables, the types of wastes considered by the Lean philosophy and the suggestions of change by the participants.

The pre-intervention simulation sub-phase was accomplished by measuring time, distance, and number of steps (dependent variables) undertaken by nurses in the access to LSE (RT, ECK, TST, and automated infusion systems (AIS)). The simulation context was used because during the research it was not possible to monitor the tasks developed by nurses in a real context. As measuring instruments, the Nokia® 6230 mobile phone chronometer was used to monitor timing performance in seconds, rounded to the unit. Sixty meters' tape Stanley PowerWinder® was used to measure the distance travelled by nurses, with data rounded to the first decimal place. The PR counted the number of steps, and the data were triangulated with the participant itself. The monitoring was performed from the point of departure (nurses' station), arrival to LSE and return to the starting point with the respective LSE.

The third pre-intervention sub-phase was completed by the suggestion of changes presented, as a proposal like determined by Lewin [40], to the Medical Director and Chief Nurse of NHDU.

The intervention phase consisted on the application of lean 5S and JIT methodologies for the reorganization of physical space, equipment location, and NHDU inventory. The tasks performed by the researcher in this phase consisted on the organization of the contents in the NHDU large cabinet, relocation, and availability of TST and AIS. The reorganization of RT, ECK, and NHDU small cabinets was performed with the help of the nurses' management and direct care team. Other human resources such as nurses' aides and the hospital carpentry services were involved to perform small changes and to construct small furniture. Stock boxes abandoned in the hospital storage were recycled and used for better storage and visual management of cabinet contents.

The post-intervention phase was divided into two sub-phases: (i) simulations, using the same methodology and equipments applied in the pre-intervention. (ii) Unstructured interviews, using the same methodology as in the pre-intervention to collect the opinion of nurses regarding the interventions made to the unit, and how this influenced their daily routines and professional practice.

The quantitative results are presented comparing the pre-intervention with the post-intervention phases, allowing a more direct comparison of the data. The IBM SPSS Statistics version 21 and Microsoft® Office® 2013 Excel version 15 software were used for the statistical analysis of data. For the statistical hypothesis tests, the parametric Student's t-test with a significance level of 0.025 (one-sided) was used, such as the nonparametric Wilcoxon W-test with the exact significance of 0.025 (one-sided) for the poorly distributed data situations [44]. Standardized response mean, calculated through MedCalc Statistical Software version 15.2.2, was used to analyze the effect size (Cohen's d) of the intervention made by the application of Lean methodologies, representing the independent variable. The qualitative results are summarized in tables with transcription of the nurses opinions collected from the interviews and the summary of the answers given by them in the questionnaires. Spaghetti diagrams and photographs are also used for better contextualization.

Attending to the literature review and the pre-intervention phase the following hypotheses are formulated:

H01: The difference of TST time of access between pre- and post-intervention equals zero.

H02: The difference of TST distance of access travelled between pre- and post-intervention equals zero.

H03: The difference of TST number of steps of access between pre- and post-intervention equals zero.

H04: The difference of AIS time of access between pre- and post-intervention equals zero.

H05: The difference of AIS distance of access travelled between pre- and post-intervention equals zero.

H06: The difference of AIS number of steps of access between pre- and post-intervention equals zero.

4. Results of the action research

Throughout Gemba Walk, twelve unstructured interviews were carried out to nurses in order to identify their difficulties in their professional daily routines and what kind of improvements they would like to implement in NHDU (**Table 1**). The collected data focused mainly on the inadequate layout and location of equipment, poor organization of clinical material in NHDU cabinets and units of patients, obstacles, restricted circulation and workspaces, frequent journeys out of NHDU to supply missing materials and equipment, and difficulty in implementing improvements because of a great resistance to change. According to these interviews only 50% of nurses knew the existence and location of TST, and only 33.3% of nurses knew the ECK existence or location. For AIS and RT, all participants were aware of them. After Lean methodologies' intervention and education, 100% of the participants were aware of all life support equipment.

In addition to the interviews, questionnaires were delivered to 12 nurses and eight were returned, representing a 67% response rate. The purpose of the questionnaire was to identify the set of difficulties felt by nurses in their daily professional life in NHDU, mainly in emergencies, monitoring and surveillance of the acute neurosurgical patient. The questionnaire made also possible to study the kind of wastes (according to Lean philosophy) the nurses identify. The collected data from questionnaires are summarized in **Table 2** that includes suggestions provided by the respondents.

A	"My greatest difficulty in NHDU is to always have to go out of the unit to look for supplies ...either because we do not have a specific location for them either it was not replaced... Medication and serums, forget it..."
B	"We should have an adequate level of stocks according to our needs and not have to always go 'out there' seek for supplies."
C	"The NHDU should be independent from all resources of Neurosurgery... Nurses and nurses' aides should be dedicated to NHDU... Stock, equipment and supplies should be replenished regularly and directly by the supply and pharmacy services."
D	"The vital signs monitors should be fixed to the wall for not taking up space in patients' desk . . . and because sometimes they drop of the desk, usually when pulled by confused patients."
E	"It's hard to work when there is not enough space to move around the patient bed without going against curtains, literally upon us, against wheelchairs and other patient's beds."
F	"There is neither space nor conditions to lift patients to an armchair or wheelchair."
G	"Patients from one bed can touch and reach things of next patients because everything is so tight and so close to each other... Patients are potentially contaminating each other ... and we ourselves have a hard time for this cross-contamination doesn't happen, I am sure it does eventually happen. "
H	"We have no space to put a RT next to the patient's units ... it is impossible to make secure ALS with the available space that we have."
I	"We usually are trained in basic life support every year, but we should also be trained in the use of RT and ALS... I have some difficulties in perceiving the location of clinical materials in the RT because there is a bad visual perception of it."
J	"I have little practice in the use of the RT, mainly the defibrillator... We should have training..."
K	"Practices adopted in NHDU goes against scientific evidences ... but it is difficult here to make whatever change we need ... some people do not understand what good practices are."
L	"The NHDU has a lack of identity and autonomy."
M	"There is a lack of standards for admittance and clearance of patients... Even the doctors and some nurses do not understand that we only have capacity for 4 patients"
N	"We cannot take any initiative to improve anything, because they fear us to take their place."
O	"They never listen to us. They do not realize, or understand, the staff who are working with them. We could make a great contribution to the better functioning of the unit."
P	"There is a huge resistance to change... There is a fear of loss or prestige transfer."

Table 1. Excerpts from interviews.

http://dx.doi.org/10.5772/64715

Difficulties with location and access to Resuscitation Trolley (RT)	
.Nothing to declare (8).	
.Suggestions:	There ought to be another RT in the ward (1).
Difficulties with the use of RT	
.Lack of experience (4); Difficulty in quickly locate contents in an intuitively manner (4); Safety seal broken without ALS manoeuvres have been performed (4); Lack of content restocking (3); Difficulty in using RT next to units because of lack of space between beds (2); Manual ventilator is incorrectly mounted (2).	
Suggestions:	Enable outer identification of drawers content for optimal access and better perception (2); Training on ALS and RT use (4); Proper RT content replacement and functioning (3).
Difficulties with access and location of the Tracheotomy Surgical Tray (TST)	
.I do not know where it is (4); I did not know it exist (4); I think it is in the ward treatment room in the medicine car or in one of the cabinets (3); Distant and time-consuming access (4).	
Suggestions:	The TST should be on NHUD and/or in the RT (8).
Difficulties with type and number of equipment in patients units	
.Broken lights (1); Monitors short wires (2); Lack of containers for hazard waste containers (3); Lack of fixation on the wall of vital signs monitors (3); Sometimes what you need is not in the patient unit (2).	
Difficulties with layout and access to equipment in patients units	
.Lack of space and difficulty in moving along the patients units (7); Difficult access (5); Material often falls on the ground, especially the suction probes (2).	
Difficulties with type and number of material in NHDU cabinets	
.Scattered material (3); Lack of material because it is not replaced and/or excess of material by wrong inventory management (7); Insufficient number of manual ventilators comparing to the number of tracheotomized in-patients (3); Lack of stethoscopes (2).	
Difficulties with provision and access to the material in NHDU cabinets	
.Constant lack of several materials (5); Stock boxes not identified (5).	
Other barriers or difficulties	
.Absence of workbench for preparation of medication(3); Poor identification of drugs (2); More critical medication should be available in the cabinets to prevent opening the RT, e.g. amiodarone, aminophylline, hydrocortisone (3); Lack of printer (3); Unnecessary travels in and out of NHDU with long distances (4); A nurse and doctor responsible for NHDU is missing (5).	
Wastes identified in the NHDU	
Production and/or procedure	Several doctors requesting the same analysis without patient observation (6); Terminally ill patients on NHDU (4); Printing documents when they are available in digital format (4); Duplicate records on paper and digital databases (3); Unnecessary procedures and bureaucracy (3).
Defect	Lack of verification of the RT (3); Damaged equipment (4); Scarce space between beds (6); Labelling of medicines are alike, sometimes mixed (2); Loss of relevant information (1).
Movement/ Transportation	Unnecessary trips to pick up unstocked material (4); Very tall cabinets (2); Bad design of the workspace (2); Patients in isolation without direct access to the corridor (3).
Time	Nurse waiting for having nurse aide available (3); Patients waiting several hours to be admitted or for bed availability (1); Delays in medical procedures and diagnostic tests (3).
Stock	Missing or excess of material on the cabinet shelves (4).
Human potential	There should be a nurse aide responsible for replacing cabinets contents in each shift (5); Unsafe nurse and nurse aide ratios (5); Nurse rehabilitation specialists should be more in touch with patients (3); Lack of training and education for health professionals (3); Absence of effective potential of employees with disability (2); Employees overworked and unmotivated to suggest changes or improvements (2).

Note. Numbers in brackets represents the frequencies of answers.

Table 2. Data Collected from Questionnaires.

According to the previous results and analysis of the interviews, questionnaires, spaghetti diagrams, value stream mapping (data not shown), and simulations, a set of suggestions were proposed by the PR to the Medical Director and Chief Nurse of NHDU (third pre-intervention sub-phase). This proposal was drawn up from the data collected attending to the Lean philosophy, the recommendations of best practices, and the standards of Portuguese regulatory institutions. The proposal considers several suggestions for amendments procedures, layout updates of the physical space, RT and NHDU cabinets content, and different locations of the clinical material and equipment. Briefly, these suggestions were the following:

1. Place suction probes supports on the wall at each bed side (accepted);
2. Place water bottles supports to ensure suction tubes washing after manipulation (accepted);
3. Place mobile IV pole with AIS mounted at each bed side (accepted);
4. Remove vital signs monitors from patients' desks and fixate them on the wall (accepted);
5. ALS and RT handling workshops for nurse training and education (accepted);
6. RT standardization (accepted);
7. Place TST at NHDU next to nurse station (accepted);
8. Reorganization of NHDU cabinets to improve contents access, variety, and identification (accepted);
9. Place drug vault at NHDU (rejected);
10. Place double air and oxygen pressure regulators at each patient unit (rejected);
11. Place manual ventilator at each unit in the presence of tracheotomized patient (rejected);
12. Organize trolley with clinical material for isolation room (rejected);
13. Eliminate one of the beds to increase circulation space (rejected).

After approval, or disapproval, of each suggestion, Lean methodologies (5S, JIT) were undertaken to ensure a better and safer work environment for patients and staff. The cabinets were reorganized into categories to cover the various patients' needs like breathing, elimination, circulation, and administration, dressings and skin integrity, feeding, individual protection equipment. Sliding frosted glasses were removed from the cabinets, and it was possible to reduce and optimize the occupied space without decreasing the amount of material, but rather increasing its variety and availability, as seen in **Figure 1** that also illustrates the post-intervention TST location. The patients' units were likewise reorganized with the inclusion of supports for suction probes, water bottles, and AIS (in mobile IV poles). Vital signs monitors were placed at a new shelve on each patient's unit and ALS, and RT workshops has been schedule for nurse training and education.

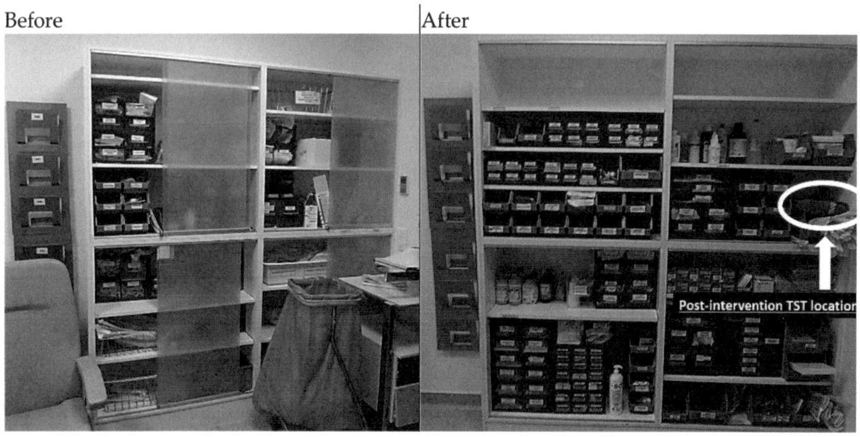

Figure 1. Supplies in NHDU large cabinet before and after Lean intervention.

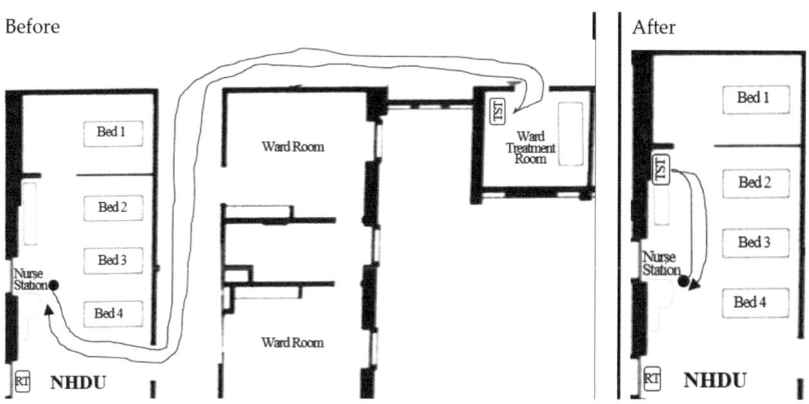

Figure 2. Spaghetti diagram for TST access before and after Lean intervention.

The presence of tracheotomized patients or at risk of being tracheotomized in NHDU is constant; therefore, the availability and accessibility to LSE, particularly ECK and TST, are of extreme importance. In the pre-intervention phase, TST was in the treatment room of neurosurgery, about 63 m (round trip) far from NHDU nurse station. Changing its location into the large cabinet inside NHDU the distance decreased to 6 m (round trip) from the nurse station. **Figure 2** represents the spaghetti diagram made before and after Lean intervention for the TST.

Table 3 shows the quantitative results obtained from the simulations of TST accessibility before and after Lean intervention. Data show the reduction of waste in time (−87.35%), distance (−90.47%), and steps (−87.12%) achieved with the application of Lean methodologies. According to Cohen's d, the effect size is large. Shapiro-Wilk normality test (data not shown) rejected the normality of the distance distribution ($p < 0.001$). So, for a one-sided significance level of 0.025, were accepted the alternative hypothesis that time ($p = 0.0017$), number of steps (p = 0.000015) and distance (p = 0.016) were statistical and significantly lower after the application of Lean methodologies.

		Time				Distance				Steps			
		A	B	Δ	Δ%	A	B	Δ	Δ%	A	B	Δ	Δ%
		Collective		Paired		Collective		Paired		Collective		Paired	
Collective and paired data	M	45.5	4.58	−40.5	−87.35	64.03	6	−58.03	−90.62	60.17	7.5	−52.5	−87.12
	Mdn	39.5	5	−34	−86.1	63	6	−57	−90.47	60.05	8	−52.5	−87.32
	SD	18.87	0.79	19.07	5.08	2.43	0	2.43	0.34	9.75	0.91	9.05	1.71
	Max	76	6	−71	−93.42	69	6	−63	−91.3	70	9	−62	−89.09
	Min	26	3	−21	−80.77	63	6	−57	−90.48	45	6	−38	−84.44
	Range	50	3	−50	−12.65	3	0	6	−0.82	25	3	−24	−4.65
t-test		95% CI			[−60.52; −20.48]								[−61.99; −43]
			t (df)		−5.2 (5)								−14.21 (5)
			p^a		0.0017								0.000015
W^b-test	Z							−2.264					
	p^a							0.016					
Effect size	Cohen's d				−2.12			−23.84					−5.8

A: Pre-intervention (n = 6). B: Post-intervention (n = 12).
[a]One-sided 0.025 significance.
[b]W-test with exact significance.

Table 3. Results from TST accessibility.

Although the ECK is correctly located in the RT, 66.7% (n = 8) of nurses were unaware of its existence or location. For ethical reasons, there was an imperative and urgent need to educate them, which was done by the PR to all nurses' team. In order to identify the difficulties of nurses in using the RT, simulations were performed. These simulations consisted in locating and accessing all RT contents, especially ECK. Through direct observation, it was found that all 12 nurses had some difficulties such as follows: safety seal breakage; retraction of safety latch; removal of back board; opening drawers by poor perception of the handle; finding and identifying critical medications and supplies; swing arm handling; and use of equipment including heart defibrillator. After the simulations, nurses justified their difficulties as a result of little practice and/or experience. An ALS and RT handling workshop intervention were scheduled to nurse's continuous education plan.

In the pre-intervention phase, the AIS were in a storeroom forcing nurses to a constant movement and transportation of about 84 m (round trip). Lean 5S and JIT methodologies determined changing AIS location into a mobile IV pole next to the patients' unit, permanently connected to electricity in order to ensure its permanent availability (**Figures 3** and **4**).

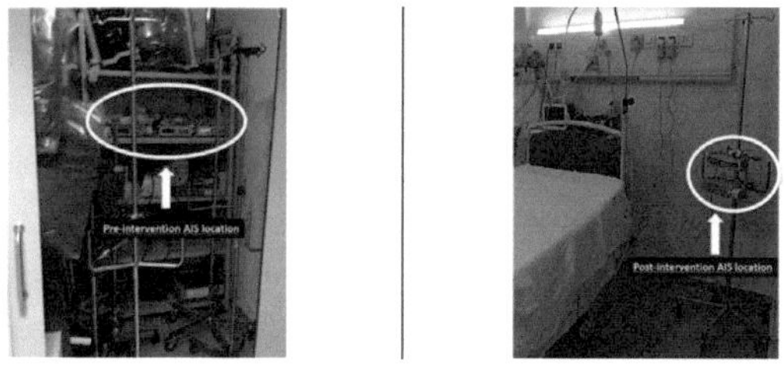

Figure 3. AIS location before and after Lean intervention.

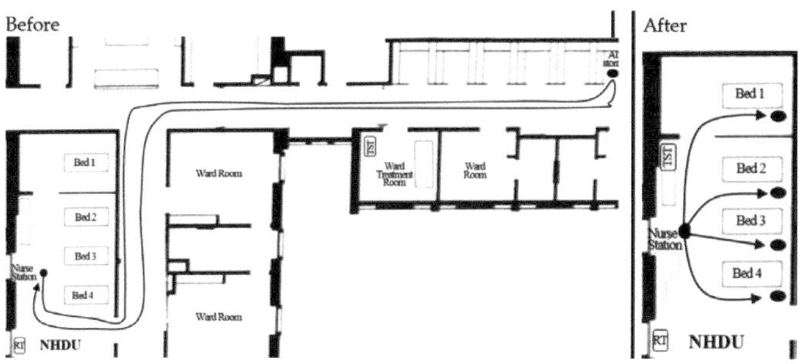

Figure 4. Spaghetti diagram for AIS access before and after Lean intervention.

		Time				Distance				Steps			
		A	B	*Δ*	*Δ*%	A	B	*Δ*	*Δ*%	A	B	*Δ*	*Δ*%
		Collective		Paired		Collective		Paired		Collective		Paired	
Collective and paired data	M	79.5	3	−76.5	−96.27	83.6	3	−80.6	−96.41	107	4.5	−102.5	−95.83
	Mdn	78.5	3	−76	−96.29	83.6	3	−80.6	−96.41	104	4.5	−99	−95.84
	SD	8.13	0.95	7.44	0.91	0	0	0	0	9.72	1.17	8.85	0.85
	Max	92	5	−87	−97.40	83.6	3	−80.6	−96.41	130	6	−124	−97.00
	Min	65	2	−63	−94.57	83.6	3	−80.6	−96.41	97	3	−94	−94.55
	Range	27	3	−24	−2.83	0	0	0	−0	33	3	−30	−2.45
t-test	95% CI			[−81.23, −71.77]									
	t (df)			−35.62 (11)									
	p[a]			5.64×10^{-13}									
W [b]-test	Z							−3.46				−3.06	
	p[a]							0.00024				0.00024	
Effect size	Cohen's d			−10.28				-5×10^{15}				−11.58	

A: Pre-intervention (*n*=6). B: Post-intervention (*n*=12).
[a]One-sided 0.025 significance.
[b]*W*-test with exact significance.

Table 4. Results from AIS accessibility.

After the intervention and application of Lean methodologies, the AIS mean access was 96.27% in time, 95.83% in steps, and 96.41% in distance lower than in the pre-intervention. The effect size is large (or very large) with d = −10.28 for time, d = −11.58 to number of steps, and d = −5×1015 to distance. For the hypothesis test, the Shapiro-Wilk normality test rejected the normality of steps (p = 0.039) and distance (this one constant) distribution. So, for a one-sided significance level of 0.025, were accepted the alternative hypothesis that time (p = 5.64×10−13), number of steps (p = 0.00024) and distance (p = 0.00024) were statistical and significantly lower after Lean methodologies application, as shown in **Table 4**.

The results of quantitative data associated to the hypothesis test, the size effect, and the improvements in accessibility to TST and AIS are summarized in **Table 5**.

Hypotheses	Statistical test	*p*-value	Size effect	Percentage variation (decrease)	Percentage variation (improvement)
H_{01}: The difference of TST time of access between pre and post-intervention equals zero.	Paired samples *t*-test	0.0017	−2.12	−87.35%	837.22%
H_{02}: The difference of TST distance of access travelled between pre and post-intervention equals zero.	Paired samples *W*-test	0.016	−23.84	−90.47%	950%
H_{03}: The difference of TST number of steps of access between pre and post-intervention equals zero.	Paired samples *t*-test	0.0000151	−5.8	−87.12%	687.46%
H_{04}: The difference of AIS time of access between pre and post-intervention equals zero.	Paired samples *t*-test	5.64×10^{-13}	−10.28	−96.27%	2733.8%
H_{05}: The difference of AIS distance of access travelled between pre and post-intervention equals zero.	Paired samples *W*-test	0.00024	-5×10^{15}	−96.41%	2686.7%
H_{06}: The difference of AIS number of steps of access between pre and post-intervention equals zero.	Paired samples *W*-test	0.00024	−11.58	−95.84%	2310%

[a] $\alpha = 0.025$ one-sided.

Table 5. Summary results from quantitative data.

5. Discussion

After all the action research phases performed, it was demonstrated that the application of Lean methodologies contributes for improving the accessibility to equipment and material that are essential to nurses' safe practice. With the application of the Lean methodologies, it is possible to provide optimized care to acute neurosurgical patients, in emergency and life support situations. Lean methodologies such as Gemba walk and spaghetti diagram made possible to identify wastes and difficulties in LSE accessibility, organization, and provision of other clinical equipment and supplies, and security issues such as potential cross-contamination provoked by exiguous work areas and architectural barriers. 5S and JIT philosophies together with interviews and questionnaires led to the development of a grounded interventional proposal for a functional and organizational harmonization of NHDU. Each suggestion on the proposal was then analyzed by medical and nurse unit managers giving deferral or refusal to certain interventions. The implementation of 5S and JIT methodologies led to the reorganization of NHDU and the allocation of the equipment closer to patients and nurses as well as to the decrease of waste, non-value-added activities and to significant improvements. These same results are argued in Carvalho et al. [45] since they defend that the layout must "reflect the need to reduce the time spent traveling" (p. 291) since "time 'lost' in travel between the various services… represents a cost to the organization in question, and that, in most cases, is not noticed or accounted for" (p. 291). For example, a nurse who searches for drugs, supplies, and equipment are doing it to serve the needs of patients, but may not notice that it can result in a waste of time, transport, handling, and human potential. But according to the Institute for Healthcare Improvement [46], if these materials were readily available when, how, and where they are needed (JIT), the time that nurses wasted looking for them would be instantly devoted to other more appropriate and critical tasks.

Through action research and the application of Lean methodologies, nurses of NHDU actually take only 10% of time, 9.37% of the distance travelled and 12.46% of the steps spent accessing TST compared to pre-intervention. The results of the intervention in AIS showed an improvement even more significant since the post-intervention access time is just 3.77% of pre-intervention time, the distance just 3.59%, and the number of steps only 4.21% compared to pre-intervention. To achieve this, nurses were educated about the location of LSE, and the need for training these nurses in ALS and RT handling was identified. Wastes and barriers that conditioned rapid access and action to acute patients were identified, reduced, or removed. Time, steps, and distance travelled accessing LSE were shortened and reduced more than half (−87.12 to −96.41%).

The same results were reached in other researches. Virginia Mason Medical Center (VMMC), in Seattle (USA), is credited to be one of the pioneers in healthcare industry to implement Lean by applying their own Virginia Mason Production System (based on TPS) [47]. Since 2001 VMMC makes efforts with the reorganization of spaces and workflows, minimizing transportation, and handling wastes, where all clinical equipment and supplies essential to care are placed in the point-of-use in UK Hereford Hospital, Lean methodologies led also to reductions of delay in nurses' response time between 40 and 93% [48]. In Scotland, from a sample of 19 critical care units, nurses available time increased from 35 to 64%, in which 32% of these units reached changes greater than 100%, supported by the program Releasing Time to Care: The Productive Ward, based on Lean and six sigma methodology [49].

In this study, there is a significant and serious lack of nurses' knowledge on the existence and location of LSE. Intervention trough education, awareness, and change of its location resulted in an improvement of 100% to TST and 200% to ECK leading to health benefits for patient's safety and quality of care. Still on the ECK and the RT, the simulation demonstrated the difficulties experienced by nurses in the use of the RT, particularly in opening it, use of drawers, location, and rapid visualization of contents. It was retrieved from this analysis that the imperative and urgent need for nurse's professional training and the need for a clearly defined intervention criteria in emergency situations. This is in line with Silich et al. [50] that also highlights that informed and trained professionals provide better care with potential reduction of adverse events, bad practices, and less waste of resources.

Catchpole [51] argues that the undesirable effects of an inadequate working environment can result in fatigue, frustration, reduced performance, and human capacity, increased risk, and adverse events. Hence, the importance that health facility managers have and the impact of their decisions on patients and staff, and "usually, it is the intermediate and elementary level manager, involved in everyday decisions, that affect the care that is actually provided to patients" [52].

6. Conclusion

This research was intended to interfere in the reality studied by solving identified problems, in an effective and participatory manner (through action), not only explain it or proposing a problem solution. The impact for practice and health services (quality indicators, safety, and satisfaction) of the Lean interventions carried out by the PR is well grounded by the results. In this research, it was verified that 66.7% of nurses were unaware of the existence or location of ECK and 50% of the TST. The education intervention resulted in an improvement of knowledge of 100% in the TST and 200% in the ECK, leading to potentially high health gains for the patient, because trained professionals provide better care with fewer mistakes. Furthermore, this research identified needs for periodic training and education on ALS and RT practice. Through Lean methodologies such as 5S, JIT, and spaghetti diagrams, it was possible to decrease time, steps, and distance travelled by nurses accessing TST and AIS between 87.12% and 96.41% and to improve this accessibility between 687.46% and 2733.8%.

These results confirm the contribution of this research to address the need of this healthcare unit to improve the care of neurosurgical acute/critically ill patients. The implementation of Lean 5S and just-in-time methodologies led to the reorganization of NHDU environment by allocating LSE closer to patients and nurses station, contributing by this way for improving the security and responsiveness of nurses' team for having more knowledge and quick access to LSE. In addition, it contributes to overcoming emergency, life support situations, and day-to-day professional life action to the needs of patients, freeing up time and availability of nurses for direct care by a work environment with less waste of time, distance, steps, handling, and setup procedures.

Although not focused in this research, for the unit and hospital management, there are potential economic and financial benefits attained from the application of Lean methodologies through the following factors: hand labor and human capital gains by reducing the time required to perform certain tasks (setup time); reduction of the "snowball" effect that leads to the accumulation of everyday work; reprocessing gains from potential reduction of costs in time of hospital internment and patient morbidity.

Besides the advantages reached with the application of the lean methodologies the research findings, however, are tempered by several shortcomings such as the unavailability of participants to collaborate with the research and resistance to change. Financial impact of the intervention was not recorded. Moreover, the results cannot be generalized; other realities can compare them and encounter similar situations that may benefit with the application of Lean methodologies in an attempt to overcome their problems.

It is expected that health professionals, especially their leaders and managers, can take some lessons from the different approaches adopted in this research and may act as a catalyst for future positive changes in all health services.

As a suggestion for future research it would be interesting to study the financial impact (time saved vs. value/hour) of the application of these lean methodologies, the impact on the quality of nurses daily professional life (satisfaction, fatigue, stress, burnout) and on emergency scenarios (LSE accessibility/availability vs. morbidity and mortality).

Author details

Ricardo Balau Esteves[1], Susana Garrido Azevedo[2*] and Francisco Proença Brójo[3]

*Address all correspondence to: sazevedo@ubi.pt

1 Health Units Management, Management and Economics Department, University of Beira Interior, Covilhã, Portugal

2 CEFAGE-UBI, Management and Economics Department, University of Beira Interior, Covilhã, Portugal

3 C-MAST – Aerospace Sciences Department, University of Beira Interior, Covilhã, Portugal

References

[1] Portuguese Directorate-General of Health. Criação e Implementação de uma Equipa de Emergência Médica Intra-hospitalar (EEMI). Circular Normative N[er] 15/DQS/DQCO ed. Lisbon: Portuguese Directorate-General of Health; 2010

[2] Ordem dos Enfermeiros. Padrões de Qualidade dos Cuidados de Enfermagem. Lisbon: Ordem dos Engenheiros; 2002

[3] Government of the Portuguese Republic. Saúde [Internet]. 2014. Available from: http://www.portugal.gov.pt/pt/os-temas/saude/a-saude.aspx [Accessed: 18 February 2014]

[4] Corvi, C. There is a waste epidemic in health care. How do you deal with it in your organization? [Internet]. 2012. Available from: http://virginiamasonblog.org/

2012/04/18/there-is-a-waste-epidemic-in-health-care-how-do-you-deal-with-it-in-your-organization [Accessed: 01 March 2014]

[5] Intensive Care Society. Standards for the care of adult patients with a temporary tracheostomy [Internet]. 2014. Available from: http://www.ics.ac.uk/EasySiteWeb/GatewayLink.aspx?alId=2212 [Accessed: 20 October 2014]

[6] Soriano-Meier, H., Forrester, P. L., Markose, S., Garza-Reyes, J. A. The role of the physical layout in the implementation of lean management initiatives. International Journal of Lean Six Sigma. 2011;2(3):254-269

[7] Womack, J. Gemba Walks. Cambridge, MA: Lean Enterprise, Inc.; 2011

[8] Imai, M. Gemba Kaizen: A Commonsense Approach to a Continuous Improvement Strategy. 2nd ed. New York, NY: McGraw-Hill; 2012

[9] Graban, M. & Swartz, J. E.. The Executive Guide to Healthcare Kaizen: Leadership for a Continuously Learning an Improving Organization. Boca Raton, FL: CRC Press, Taylor & Francis Group; 2014

[10] Meisel, R. M., Babb, S. J., Marsh, S. F., Schlichting, J. P. The Executive Guide to Understanding and Implementing Lean Six Sigma: The Financial Impact. Milwaukee, WI: American Society for Quality, Quality Press; 2007

[11] Locher, D. A. Value Stream Mapping for Lean Development: A How-To Guide for Streamlining Time to Market. New York, NY: Productivity Press, Taylor & Francis Group; 2008

[12] Jackson, T. L., editors. 5S for Healthcare. New York, NY: Rona Consulting Group & Productivity Press; 2009

[13] Smart, N. J. Lean Biomanufacturing: Creating Value through Innovative Bioprocessing Approaches. Cambridge: Woodhead Publishing Limited; 2013

[14] Womack, J. P., Jones, D. T. Lean Thinking: Banish Waste and Create Wealth in Your Corporation. New York, NY: Free Press, 2003

[15] Bialek, R., Duffy, G. L. & Moran, J. W., editors. The Public Health Quality Improvement Handbook. Milwaukee, WI: ASQ Quality Press; 2009

[16] Krafcik, J.. Triumph of the lean production system. Sloan Management Review. 1988;30(1):41-52

[17] Womack, J. P., Jones, D. T. & Roos, D.. The Machine That Changed the World. New York, NY: Rawson Associates; Collier Macmillan Canada; 1990

[18] Treville, S. & Antonakis, J. Could Lean production job design, be intrinsically motivating? Contextual, configurational, and levels-of-analysis issues. Journal of Operations Management. 2006;24(2):99-123

[19] Anand, G. & Kodali, R. Development of a framework for implementation of Lean manufacturing systems. International Journal of Management Practice. 2008;4(1):95-116

[20] Brown, K. & Mitchell, T. A comparison of just-in-time and batch manufacturing: the role of performance obstacles. Academy of Management, Journal. 1991;34(4):906-917

[21] Gurumurthy, A. & Kodali, R. Application of benchmarking for assessing the Lean manufacturing implementation. Benchmarking: An International Journal. 2009;16(2):274-308

[22] Demeter, K. & Matyusz, Z. The impact of lean practices on inventory turnover. International Journal of Production Economics. 2011;133(1):154-163

[23] Stewart, D. & Grout, J. The human side of mistake-proofing. Production and Operations Management. 2001;10(4):440-459

[24] Melton T. The benefits of Lean manufacturing what Lean thinking has to offer the process industries. Chemical Engineering Research and Design. 2005;83(6):662-673

[25] Aheme, J. & Whelton, J., editors. Applying Lean in Healthcare: A Collection of International Case Studies. Boca Raton, FL: CRC Press – Taylor & Francis Group; 2010

[26] Silva, R. C. Seis Sigma na Avaliação Externa da Qualidade em Laboratórios Clínicos [dissertation]. Lisbon: University of Lisbon; 2013. Available from: http://hdl.handle.net/10362/10155

[27] Valente, R. P., Esteves, M. & Padilha, J. A metodologia Lean na área hospitalar - a gestão da qualidade enquanto factor de melhoria contínua e humanização do esforço de racionalização dos recursos. In: III Seminário de I&DT; 6-7 December 2012; Portalegre. Portalegre: C3i – Centro Interdisciplinar de Investigação e Inovação do Instituto Politécnico de Portalegre in Portalegre; 2012

[28] Araújo, M. M. A. Lean nos Serviços de Saúde [dissertation]. Porto: University of Porto; 2009. Available from: http://hdl.handle.net/10216/59676

[29] Pinto, J. M. M. Kaizen nas Unidades Hospitalares [dissertation]. Porto: University of Porto; 2008. Available from: http://hdl.handle.net/10216/57973

[30] Simões, F. M. C. Lean Healthcare – O conceito Lean aplicado à realidade dos serviços de saúde [dissertation]. Aveiro: University of Aveiro; 2009. Available from: http://hdl.handle.net/10773/1729

[31] Carvalho, J. L. A. Reengenharia de Processos na Indústria Farmacêutica [dissertation]. Lisbon: University Nova of Lisbon; 2010. Available from: http://hdl.handle.net/10362/4029

[32] Libano, L. E. M. Análise Económica da Introdução do Sistema de Melhoria Contínua Kaizen no Bloco Operatório do Centro Hospitalar do Porto [thesis]. Braga: University

of Minho; 2012. Available from: http://repositorium.sdum.uminho.pt/bitstream/1822/23481/1/Lu%C3%ADsa%20Emanuela%20Martins%20Libano.pdf

[33] Matos, I. A. Aplicação de técnicas Lean Services no bloco operatório de um hospital [dissertation]. Guimarães: University of Minho; 2011. Available from: http://hdl.handle.net/1822/16321

[34] Paula, P. S. A contribuição da implementação dos 5S para a melhoria contínua da qualidade num serviço de imagiologia - o estudo de caso no HFF [dissertation]. Porto: University Fernando Pessoa; 2008. Available from: http://hdl.handle.net/10284/1431

[35] Dias, S. M. Implementação da metodologia Lean Seis-Sigma – O caso do Serviço de Oftalmologia dos Hospitais da Universidade de Coimbra [dissertation]. Coimbra: University of Coimbra; 2011. Available from: http://hdl.handle.net/10316/17667

[36] Resende, M. O. Melhoria de Processos Hospitalares através de ferramentas Lean: Aplicação ao serviço de Imagiologia no Centro Hospitalar Entre Douro e Vouga [dissertation]. Porto: University of Porto; 2010. Available from: http://repositorio-aberto.up.pt/bitstream/10216/59520/1/000145447.pdf

[37] Ribeiro, A. C. A implementação da filosofia Lean na gestão dos serviços de saúde: o caso dos centros de saúde da região norte [dissertation]. Porto: University of Porto; 2013. Available from: http://hdl.handle.net/10216/69710

[38] Guimarães, M. C. M. Lean thinking in Healthcare services - learning from case studies [thesis]. Lisbon: Lisbon University Institut ; 2013. Available from: http://hdl.handle.net/10071/6183

[39] Luzes, C. S. A. Implementação da Filosofia Lean na Gestão dos Serviços de Saúde: O Caso Português [dissertation]. Porto: Polytechnic Institut of Porto; 2013. Available from: http://www.fep.up.pt/docentes/fontes/FCTEGE2008/Publicacoes/D17.pdf

[40] Lewin, K. Action research and minority problems. Journal of Social Issues. 1946;2(4): 34-46

[41] Given, L. M. The Sage Encyclopedia of Qualitative Research Methods. Thousand Oaks, CA: SAGE Publications, Inc.; 2008

[42] Streubert, H. J. & Carpenter, D. R.. Investigação Qualitativa em Enfermagem: Avançando o imperativo humanista. Loures: Lusociência; 2002

[43] Yin, R. K. Qualitative Research from Start to Finish. New York, NY: The Guilford Press; 2011

[44] Mehta, C. R., Patel, N. R. IBM SPSS Exact Tests. Cambridge, MA: IBM Corporation; 2012

[45] Carvalho, J. C., & Ramos, T. Logística na Saúde. Lisbon: Edições Sílabo, Lda; 2009

[46] Institute for Healthcare Improvement. Going Lean in Health Care. Cambridge, MA: Institute for Healthcare Improvement; 2005

[47] Virginia Mason Medical Center. Virginia Mason Production System - Fast Facts [Internet]. 2015. Available from: https://www.virginiamason.org/workfiles/pdfdocs/press/vmps_ fastfacts.pdf [Accessed: 20 January 2015]

[48] Institute for Innovation and Improvement. Reducing Turnaround Times in Pathology using Lean Thinking [Internet]. 2006. Available from: http://www.institute.nhs.uk/images//documents/Quality_and_value/Lean/herefordpathologycasestudyinstitutela-testsept2006.pdf [Accessed: 05 May 2014]

[49] Healthcare Improvement Scotland. Releasing Time to Care: Making your Priorities Possible [Internet]. 2013. Available from: http://www.healthcareimprovementscot-land.org/our_work/person-centred_care/releasing_time_to_care.aspx [Accessed: 20 January 2015]

[50] Silich, S. J., Wetz, R. V., Riebling, N., Coleman, C., Khoueiry, G., Nidal, A. R., Szerszen, A. Using six sigma methodology to reduce patient transfer times from floor to critical-care beds. Journal for Healthcare Quality. 2012;34(1):44-54

[51] Catchpole, K. Human Factors. In: S. S. Panesar, A. Carson-Stevens, S. A. Salvilla & A. Sheikh, editors. Patient Safety and Healthcare Improvement at a Glance. Chichester: John Wiley & Sons, Ltd; 2014. p. 8-11

[52] Crisp, L. N., Berwick, D., Kickbusch, I., Bos, W., Antunes, J. L., Barros, P. P., & Soares, J. Um Futuro para a Saúde – todos temos um papel a desempenhar. Lisbon: Foundation Calouste Gulbenkian; 2014

Chapter 4

Iteration Algorithms in Markov Decision Processes with State-Action-Dependent Discount Factors and Unbounded Costs

Fernando Luque-Vásquez and
J. Adolfo Minjárez-Sosa

Additional information is available at the end of the chapter

http://dx.doi.org/10.5772/65044

Abstract

This chapter concerns discrete time Markov decision processes under a discounted optimality criterion with state-action-dependent discount factors, possibly unbounded costs, and noncompact admissible action sets. Under mild conditions, we show the existence of stationary optimal policies and we introduce the value iteration and the policy iteration algorithms to approximate the value function.

Keywords: discounted optimality, non-constant discount factor, value iteration, policy iteration, Markov decision processes

AMS 2010 subject classifications: 93E10, 90C40

1. Introduction

In this chapter we study Markov decision processes (MDPs) with Borel state and action spaces under a discounted criterion with state-action–dependent discount factors, possibly unbounded-ed costs and noncompact admissible action sets. That is, we consider discount factors of the form

$$\alpha(x_n, a_n), \tag{1}$$

where x_n and a_n are the state and the action at time n, respectively, playing the following role during the evolution of the system. At the initial state x_0, the controller chooses an action a_0 and a cost $c(x_0,a_0)$ is incurred. Then the system moves to a new state x_1 according to a transition law. Once the system is in state x_1 the controller selects an action a_1 and incurs a discounted cost $\alpha(x_0,a_0)c(x_1,a_1)$. Next the system moves to a state x_2 and the process is repeated. In general, for the stage $n \geq 1$, the controller incurs the discounted cost

$$\prod_{k=0}^{n-1} \alpha(x_k, a_k) c(x_n, a_n), \tag{2}$$

and our objective is to show the existence of stationary optimal control policies under the corresponding performance index, as well as to introduce approximation algorithms, namely, value iteration and policy iteration.

In the scenario of assuming a constant discount factor, the discounted optimality criterion in stochastic decision problems is the best understood of all performance indices, and it is widely accepted in several application problems (see, e.g., [1–3] and reference there in). However, such assumption might be strong or unrealistic in some economic and financial models. Indeed, in these problems the discount factors are typically functions of the interest rates, which in turn depend on the amount of currency and the decision-makers actions. Hence, we have state-action–dependent discount factors, and it is indeed these kinds of situations we are dealing with.

MDPs with non constant discount factors have been studied under different approaches (see, e.g., [4–8]). In particular, our work is a sequel to [8] where is studied the control problem with state-dependent discount factor. In addition, randomized discounted criteria have been analyzed in [9–12] where the discount factor is modeled as a stochastic process independent of the state-action pairs.

Specifically, in this chapter we study control models with state-action-dependent discount factors, focusing mainly on introducing approximation algorithms for the optimal value function (value iteration and policy iteration). Furthermore, an important feature in this work is that there is no compactness assumption on the sets of admissible actions neither continuity conditions on the cost, which, in most of the papers on MDPs, are needed to show the existence of measurable selectors and continuity or semi-continuity of the minima function. Indeed, in contrast to the previously cited references, in this work, we assume that the cost and discount factor functions satisfy the $\mathcal{K}$-inf-compactness condition introduced in [13]. Then, we use a generalization of Berge's Theorem, given in [13], to prove the existence of measurable selectors. To the best of our knowledge there are no works dealing with MDPs in the context presented in this chapter.

The remainder of the chapter is organized as follows. Section 2 contains the description of the Markov decision model and the optimality criterion. In Section 3 we introduce the assumptions on the model and we prove the convergence of the value iteration algorithm (Theorem 3.5). In Section 4 we define the policy iteration algorithm and the convergence is stated in Theorem 4.1.

Notation. Throughout the paper we shall use the following notation. Given a *Borel space* S— that is, a Borel subset of a complete separable metric space — $\mathcal{B}(S)$ denotes the Borel σ-algebra and "measurability" always means measurability with respect to $\mathcal{B}(S)$. Given two Borel spaces S and S', a *stochastic kernel* $\varphi(\cdot \mid \cdot)$ on S given S' is a function such that $\varphi(\cdot \mid s')$ is a probability measure on S for each $s' \in S'$, and $\varphi(B \mid \cdot)$ is a measurable function on S' for each $B \in \mathcal{B}(S)$. Moreover, $\mathbb{N}$ ($\mathbb{N}_0$) denotes the positive (nonnegative) integers numbers. Finally, $L(S)$ stands for the class of lower semicontinuous functions on S bounded below and $L_+(S)$ denotes the subclass of nonnegative functions in $L(S)$.

2. Markov decision processes

Markov control model. Let

$$\mathcal{M} := \left(X, A, \{A(x) \subset A \mid x \in X\}, Q, \alpha, c\right) \tag{3}$$

be a discrete-time Markov control model with state-action-dependent discount factors satisfying the following conditions. The state space X and the action or control space A are Borel spaces. For each state $x \in X$, $A(x)$ is a nonempty Borel subset of A denoting the set of admissible controls when the system is in state x. We denote by $\mathbb{K}$ the graph of the multifunction $x \mapsto A(x)$, that is,

$$\mathbb{K} = \{(x,a) : x \in X, a \in A(x)\} \tag{4}$$

which is assumed to be a Borel subset of the Cartesian product of X and A. The transition law $Q(\cdot \mid \cdot)$ is a stochastic kernel on X given $\mathbb{K}$. Finally, $\alpha : \mathbb{K} \to (0,1)$ and $c : \mathbb{K} \to (0,\infty)$ are measurable functions representing the discount factor and the cost-per-stage, respectively, when the system is in state $x \in X$ and the action $a \in A(x)$ is selected.

The model $\mathcal{M}$ represents a controlled stochastic system and has the following interpretation. Suppose that at time $n \in \mathbb{N}_0$ the system is in the state $x_n = x \in X$. Then, possibly taking into account the history of the system, the controller selects an action $a_n = a \in A(x)$, and a discount factor $\alpha(x, a)$ is imposed. As a consequence of this the following happens:

1. A cost $c(x,a)$ is incurred;
2. The system visits a new state $x_{n+1} = x' \in X$ according to the transition law

$$Q(B \mid x,a) := \Pr[x_{n+1} \in B \mid x_n = x, a_n = a], \quad B \in \mathcal{B}(X). \tag{5}$$

Once the transition to state x' occurs, the process is repeated.

Typically, in many applications, the evolution of the system is determined by stochastic difference equations of the form

$$x_{n+1} = F(x_n, a_n, \xi_n), \, n \in \mathbb{N}_0, \tag{6}$$

where $\{\xi_n\}$ is a sequence of independent and identically distributed random variables with values in some Borel space S, independent of the initial state x_0, and $F: X \times A \times S \rightarrow X$ is a given measurable function. In this case, if θ denotes the common distribution of ξ_n, that is

$$\theta(D) := P[\xi_n \in D], \ D \in \mathcal{B}(S), \, n \in \mathbb{N}_0, \tag{7}$$

then the transition kernel can be written as

$$\begin{aligned} Q(B \mid x,a) &= \Pr[F(x_n, a_n, \xi_n) \in B \mid x_n = x, a_n = a] \\ &= \theta\{s \in S : F(x,a,s) \in B\} \\ &= \int_S 1_B[F(x,a,s)]\theta(ds), \quad B \in \mathcal{B}(X), (x,a) \in \mathbb{K}, \end{aligned} \tag{8}$$

where $1_B(\cdot)$ represents the indicator function of the set B.

Control policies. The actions applied by the controller are chosen by mean of rules known as control policies defined as follows. Let $\mathbb{H}_0 := X$ and $\mathbb{H}_n := \mathbb{K}^n \times X, n \geq 1$ be the spaces of admissible histories up to time n. A generic element of $\mathbb{H}_n$ is written as $h_n = (x_0, a_0, \ldots, x_{n-1}, a_{n-1}, x_n)$.

Definition 2.1 *A control policy (randomized, history-dependent) is a sequence $\pi = \{\pi_n\}$ of stochastic kernels π_n on A given $\mathbb{H}_n$ such that $\pi_n(A(x_n) \mid h_n) = 1$, for all $h_n \in \mathbb{H}_n$, $n \in \mathbb{N}_0$.*

We denote by Π the set of all control policies.

Let $\mathbb{F}$ be the set of measurable selectors, that is, $\mathbb{F}$ is the set of measurable function $f: X \rightarrow A$ such that $f(x) \in A(x)$ for all $x \in X$.

Definition 2.2 *A control policy $\pi = \{\pi_n\}$ is said to be:*

a. *deterministic if there exists a sequence $\{g_n\}$ of measurable functions $g_n: \mathbb{H}_n \rightarrow A$ such that*

$$\pi_n(C \mid h_n) = 1_C\left[g_n(h_n)\right], \ \forall h_n \in \mathbb{H}_n, n \in \mathbb{N}_0, C \in \mathcal{B}(A); \tag{9}$$

b. *a Markov control policy if there exists a sequence* $\{f_n\}$ *of functions* $f_n \in \mathbb{F}$ *such that*

$$\pi_n(C \mid h_n) = 1_C\left[f_n(x_n)\right], \ \forall h_n \in \mathbb{H}_n, n \in \mathbb{N}_0, C \in \mathcal{B}(A). \tag{10}$$

In addition

c. *A Markov control policy is stationary if there exists* $f \in \mathbb{F}$ *such that* $f_n = f$ *for all* $n \in \mathbb{N}_0$.

If necessary, see for example [1–3, 14–16] for further information on those policies.

Observe that a Markov policy π is identified with the sequence $\{f_n\}$, and we denote $\pi = \{f_n\}$. In this case, the control applied at time n is $a_n = f_n(x_n) \in A(x_n)$. In particular, a stationary policy is identified with the function $f \in \mathbb{F}$, and following a standard convention we denote by $\mathbb{F}$ the set of all stationary control policies.

To ease the notation, for each $f \in \mathbb{F}$ and $x \in X$, we write

$$\begin{aligned} c(x, f) \quad &: \ = c(x, f(x)), \\ Q(\cdot \mid x, f) \quad &: \ = Q(\cdot \mid x, f(x)), \end{aligned} \tag{11}$$

and

$$\alpha(x, f) := \alpha(x, f(x)). \tag{12}$$

The underlying probability space. Let $(\Omega, \mathcal{F})$ be the canonical measurable space consisting of the sample space $\Omega = \mathbb{K}^{\infty} := \mathbb{K} \times \mathbb{K} \times \cdots$ and its product σ −algebra $\mathcal{F}$. Then, under standard arguments (see, e.g., [1, 14]) for each $\pi \in \Pi$ and initial state $x \in X$, there exists a probability measure P_x^{π} on $(\Omega, \mathcal{F})$ such that, for all $h_n \in \mathbb{H}_n$, $a_n \in A(x_n)$, $n \in \mathbb{N}_0$, $C \in \mathcal{B}(A)$, and $B \in \mathcal{B}(X)$,

$$\begin{aligned} P_x^{\pi}\left[x_0 = x\right] &= 1; \\ P_x^{\pi}\left[a_n \in C \mid h_n\right] &= \pi_n(C \mid h_n); \end{aligned} \tag{13}$$

and the Markov-like property is satisfied

$$P_x^{\pi}\left[x_{n+1} \in B \mid h_n, a_n\right] = Q(B \mid x_n, a_n). \tag{14}$$

The stochastic process $\left(\Omega, \mathcal{F}, P_x^\pi, \{x_n\}\right)$ is called Markov decision process.

Optimality criterion. We assume that the costs are discounted in a multiplicative discounted rate. That is, a cost C incurred at stage n is equivalent to a cost $C\Gamma_n$ at time 0, where

$$\Gamma_n := \prod_{k=0}^{n-1} \alpha(x_k, a_k) \text{ if } n \geq 1, \text{ and } \Gamma_0 = 1. \tag{15}$$

In this sense, when using a policy $\pi \in \Pi$, given the initial state $x_0 = x$, we define the total expected discounted cost (with state-action–dependent discount factors) as

$$V(\pi, x) := E_x^\pi \left[\sum_{n=0}^{\infty} \Gamma_n c(x_n, a_n) \right], \tag{16}$$

where E_x^π denotes the expectation operator with respect to the probability measure P_x^π induced by the policy π, given $x_0 = x$.

The optimal control problem associated to the control model $\mathcal{M}$, is then to find an optimal policy $\pi^* \in \Pi$ such that $V(\pi^*, x) = V(x)$ for all $x \in X$, where

$$V(x) := \inf_{\pi \in \Pi} V(\pi, x) \tag{17}$$

is the optimal value function (see [10]).

3. The value iteration algorithm

In this section we give conditions on the model that imply: (i) the convergence of the value iteration algorithm; (ii) the value function V is a solution of the corresponding optimality equation; and (iii) the existence of stationary optimal policies. In order to guarantee that $V(x)$ is finite for each initial state x we suppose the following.

Assumption 3.1. *There exists $\pi_0 \in \Pi$ such that for all $x \in X, V(\pi_0, x) < \infty$.*

At the end of Section 4 we give sufficient conditions for Assumption 3.1. We also require continuity and (inf-) compactness conditions to ensure the existence of "measurable minimizers." The following definition was introduced in [13].

Definition 3.2. *A function $u: \mathbb{K} \to \mathbb{R}$ is said to be $\mathcal{K}$-inf-compact on $\mathbb{K}$ if for each compact subset K of X and $r \in \mathbb{R}$, the set*

$$\{(x,a) \in Gr_K(A) : u(x,a) \leq r\} \tag{18}$$

is a compact subset of $X \times A$, *where* $Gr_K(A) := \{(x,a) : x \in K, a \in A(x)\}$.

Assumption 3.3. *(a) The one-stage cost* c *and the discount factor* α *are* $\mathcal{K}$*-inf-compact functions on* $\mathbb{K}$. *In addition,* c *is nonnegative.*

(b) The transition law Q is weakly continuous, that is, the mapping

$$(x,a) \to \int_X u(y) Q(dy \mid x,a) \tag{19}$$

is continuous for each bounded and continuous function on X.

For each measurable function u on $X, x \in X$, and $f \in \mathbb{F}$, we define the operators

$$Tu(x) := \inf_{a \in A(x)} \left\{ c(x,a) + \alpha(x,a) \int_X u(y) Q(dy \mid x,a) \right\} \tag{20}$$

and

$$T_f u(x) := c(x,f) + \alpha(x,f) \int_X u(y) Q(dy \mid x,f). \tag{21}$$

A consequence of Assumption 3.3 is the following.

Lemma 3.4. *Let* u *be a function in* $L_+(X)$. *If Assumption 3.3 holds then the function* $v : \mathbb{K} \to \mathbb{R}$ *defined by*

$$v(x,a) := c(x,a) + \alpha(x,a) \int_X u(y) Q(dy \mid x,a) \tag{22}$$

is $\mathcal{K}$*-inf-compact on* $\mathbb{K}$

Proof. First note that by the $\mathcal{K}$-inf-compactness hypothesis $c(\cdot,\cdot)$ and $\alpha(\cdot,\cdot)$ are l.s.c on $Gr_K(A)$ for each compact subset K of X. Then, since α and u are nonnegative functions, from Assumption 3.3 we have that $v(\cdot,\cdot)$ is l.s.c on $Gr_K(A)$. Thus, for each $r \in \mathbb{R}$, the set

$$\{(x,a) \in Gr_K(A) : v(x,a) \leq r\} \tag{23}$$

is a closed subset of the compact set $\{(x,a) \in Gr_K(A): c(x,a) \leq r\}$. Then, v is $\mathcal{K}$-inf-compact on $\mathbb{K}$.

Observe that the operator T is monotone in the sense that if $v \geq u$ then $Tv \geq Tu$. In addition, from Assumption 3.3 and ([13], Theorem 3.3), we have that T maps $L_+(X)$ into itself. Furthermore, there exists $\tilde{f} \in \mathbb{F}$ such that

$$Tu(x) = T_{\tilde{f}}u(x), \quad x \in X. \tag{24}$$

To state our first result we define the sequence $\{v_n\} \subset L_+(X)$ of value iteration functions as:

$$\begin{aligned} v_0 &\equiv 0; \\ v_n(x) &= Tv_{n-1}(x), \quad x \in X. \end{aligned} \tag{25}$$

Since T is monotone, note that $\{v_n\}$ is a nondecreasing sequence.

Theorem 3.5. *Suppose that Assumptions 3.1 and 3.3 hold. Then*

a. $v_n \nearrow V$.

b. *V is the minimal solution in $L_+(X)$ of the Optimality Equation, i.e.,*

$$V(x) = TV(x) = \inf_{a \in A(x)} \left\{ c(x,a) + \alpha(x,a) \int_X V(y) Q(dy \mid x,a) \right\}. \tag{26}$$

c. *There exists a stationary policy $f^* \in \mathbb{F}$ such that, for all $x \in X, V(x) = T_{f^*}V(x)$, that is*

$$V(x) = c(x,f^*) + \alpha(x,f^*) \int_X u(y) Q(dy \mid x, f^*), \tag{27}$$

and f^* is an optimal policy.

Proof. Since $\{v_n\}$ is nondecreasing, there exists $v \in L_+(X)$ such that $v_n \nearrow v$. Hence, from Monotone Convergence Theorem, ([13], Lemmas 2.2, 2.3), and ([1], Lemma 4.2.4), we obtain, for each $x \in X$, $v_n(x) = Tv_{n-1}(x) \to Tv(x)$, as $n \to \infty$, which, in turn implies

$$Tv = v. \tag{28}$$

Therefore, to get (a)-(b) we need to prove that $v = V$. To this end, observe that for all $x \in X$ and $\pi \in \Pi$

$$v_n(x) \leq \int_A c(x,a)\pi(da \mid x) + \int_A \alpha(x,a) \int_X v_{n-1}(x_1) Q(dx_1 \mid x,a)\pi(da \mid x). \tag{29}$$

Then, iterating (29) we obtain

$$v_n(x) \leq V_n(\pi, x), \ \ n \in \mathbb{N}, \tag{30}$$

where

$$V_n(\pi, x) = E_x^{\pi}\left[\sum_{t=0}^{n-1} \Gamma_t c(x_t, a_t)\right], \tag{31}$$

is the n −stage discounted cost V_n. Then, letting $n \to \infty$ we get $v(x) \leq V(\pi,x)$, for all $\pi \in \Pi$ and $x \in X$. Thus,

$$v(x) \leq V(x), \ \ x \in X. \tag{32}$$

On the other hand, from (28) and (24), let $f \in \mathbb{F}$ such that $v(x) = T_f v(x)$, $x \in X$. Iterating this equation, we have (see (31))

$$\begin{aligned} v(x) &= E_x^f\left[c(x,f) + \sum_{t=1}^{n-1} \prod_{k=0}^{t-1} \alpha(x_k, f) c(x_t, f)\right] \\ &\quad + E_x^f\left[\prod_{k=0}^{n-1} \alpha(x_k, f) v(x_n)\right] \\ &\geq V_n(f, x). \end{aligned} \tag{33}$$

Hence, letting $n \to \infty$,

$$v(x) \geq V(f, x) \geq V(x), \ \ x \in X. \tag{34}$$

Combining (32) and (34) we get $v = V$.

Now, let $u \in L_+(X)$ be an arbitrary solution of the optimality equation, that is, $Tu = u$. Then, applying the arguments in the proof of (34) with u instead of v we conclude that $u \geq V$. That is, V is minimal in $L_+(X)$.

Part (c) follows from (b) and ([13], Theorem 3.3). Indeed, there exists a stationary policy $f^* \in \mathbb{F}$ such that $V(x) = T_{f^*}V(x), x \in X$. Then, iteration of this equation yields $V(x) = V(f^*,x)$, which implies that f^* is optimal.

4. Policy iteration algorithm

In Theorem 3.5 is established an approximation algorithm for the value function V by means of the sequence of the value iteration functions $\{v_n\}$. In this case the sequence $\{v_n\}$ increase to V and it is defined recursively. Now we present the well-known policy iteration algorithm which provides a decreasing approximation to V in the set of the control policies.

To define the algorithm, first observe that from the Markov property (14) and applying properties of conditional expectation, for any stationary policy $f \in \mathbb{F}$ and $x \in X$, the corresponding cost $V(f,x)$ satisfies

$$\begin{aligned} V(f,x) &= c(x,f) + \alpha(x,f)E_x^f\left[\sum_{t=1}^{\infty}\prod_{k=0}^{t-1}\alpha(x_k,f)c(x_t,f)\right] \\ &= c(x,f) + \alpha(x,f)\int_X E^f\left[c(x_1,f) + \sum_{t=2}^{\infty}\prod_{k=0}^{t-1}\alpha(x_k,f)c(x_t,f) \mid x_1 = y\right]Q(dy \mid x,f) \\ &= c(x,f) + \alpha(x,f)\int_X V(f,y)Q(dy \mid x,f) = T_f V(f,x), \quad x \in X. \end{aligned} \tag{35}$$

Let $f_0 \in \mathbb{F}$ be a stationary policy with a finite valued cost $w_0(\cdot) := V(f_0,\cdot) \in L_+(X)$. Then, from (35),

$$\begin{aligned} w_0(x) &= c(x,f_0) + \alpha(x,f_0)\int_X w_0(y)Q(dy \mid x,f_0) \\ &= T_{f_0}w_0(x), \quad x \in X. \end{aligned} \tag{36}$$

Now, let $f_1 \in \mathbb{F}$ be such that

$$Tw_0(x) = T_{f_1}w_0(x), \tag{37}$$

and define $w_1(\cdot) = V(f_1,\cdot)$.

In general, we define a sequence $\{w_n\}$ in $L_+(X)$ as follows. Given $f_n \in \mathbb{F}$, compute $w_n(\cdot) := V(f_n,\cdot) \in L_+(X)$ Next, let $f_{n+1} \in \mathbb{F}$ be such that

http://dx.doi.org/10.5772/65044

$$T_{f_{n+1}} w_n(x) = Tw_n(x), \quad x \in X, \tag{38}$$

that is,

$$\begin{aligned} T_{f_{n+1}} w_n(x) &= c(x, f_{n+1}) + \alpha(x, f_{n+1}) \int_X w_n(y) Q(dy|x, f_{n+1}) \\ &= \min_{a \in A(x)} \left\{ c(x,a) + \alpha(x,a) \int_X w_n(y) Q(dy|x,a) \right\} \\ &= Tw_n(x), \quad x \in X. \end{aligned} \tag{39}$$

Then we define $w_{n+1}(\cdot) = V(f_{n+1}, \cdot)$

Theorem 4.1. *Under Assumptions 3.1 and 3.3, there exists a measurable nonnegative function* $w \geq V$ *such that* $w_n \searrow w$, *and* $Tw = w$. *Moreover, if w satisfies*

$$\lim_{n \to \infty} E_x^\pi [\Gamma_n w(x_n)] = 0 \quad \forall \pi \in \Pi, \ x \in X, \tag{40}$$

then $w = V$.

To prove the Theorem 4.1 we need the following result.

Lemma 4.2. *Under Assumption 3.3, if* $u: X \to \mathbb{R}$ *is a measurable function such that* Tu *is well defined,* $u \leq Tu$, *and*

$$\lim_{n \to \infty} E_x^\pi \left[\Gamma_n u(x_n) \right] = 0 \quad \forall \pi \in \Pi, x \in X, \tag{41}$$

then $u \leq V$.

Proof. From the Markov property (14), for each $\pi \in \Pi$ and $x \in X$,

$$E_x^\pi \left[\Gamma_{n+1} u(x_{n+1}) \,|\, h_n, a_n \right] = \Gamma_{n+1} \int_X u(y) Q(dy \,|\, x_n, a_n) \tag{42}$$

$$= \Gamma_n \left[c(x_n, a_n) + \alpha(x_n, a_n) \int_X u(y) Q(dy \,|\, x_n, a_n) - c(x_n, a_n) \right] \tag{43}$$

$$\geq \Gamma_n \inf_{a \in A(x_n)} \left[c(x_n, a) + \alpha(x_n, a) \int_X u(y) Q(dy \,|\, x_n, a) \right] - \Gamma_n c(x_n, a_n) \tag{44}$$

$$= \Gamma_n Tu(x_n) - \Gamma_n c(x_n, a_n) \geq \Gamma_n u(x_n) - \Gamma_n c(x_n, a_n), \tag{45}$$

which, in turn implies

$$\Gamma_n c(x_n,a_n) \geq E_x^\pi\left[\Gamma_n u(x_n) - \Gamma_{n+1} u(x_{n+1}) \mid h_n,a_n\right]. \tag{46}$$

Therefore, for all $k \in \mathbb{N}$ (see (31)),

$$V_k(\pi,x) = E_x^\pi \sum_{n=0}^{k-1} \Gamma_n c(x_n,a_n) \geq u(x) - E_x^\pi\left[\Gamma_k u(x_k)\right]. \tag{47}$$

Finally, letting $k \to \infty$, (41) yields $V(\pi,x) \geq u(x)$, and since π is arbitrary we obtain $V(x) \geq u(x)$.

Proof of Theorem 4.1. According to Lemma 4.2, it is sufficient to show the existence of a function $w \geq V$ such that $w_n \searrow w$ and $Tw = w$. To this end, from (36)–(38),

$$\begin{aligned} w_0(x) &\geq \min_{a\in A(x)} \left\{ c(x,a) + \alpha(x,a) \int_X w_0(y) Q(dy|x,a) \right\} = T_{f_1} w_0(x) \\ &= c(x,f_1) + \alpha(x,f_1) \int_X w_0(y) Q(dy|x,f_1). \end{aligned} \tag{48}$$

Iterating this inequality, a straightforward calculation as in (34) shows that

$$w_0(x) \geq V(f_1,x) = w_1(x), \quad x \in X. \tag{49}$$

In general, similar arguments yield

$$w_n \geq Tw_n \geq w_{n+1}, \quad n \in \mathbb{N}. \tag{50}$$

Therefore, there exists a nonnegative measurable function w such that $w_n \searrow w$. In addition, since $w_n \geq V\ \forall n \in \mathbb{N}_0$, $\forall n \in \mathbb{N}_0$, we have $w \geq V$. Next, letting $n \to \infty$ in (47) and applying ([17], Lemma 3.3), we obtain $w \geq Tw \geq w$, which implies $w = Tw$.

4.1. Sufficient conditions for Assumption 3.1 and (40)

An obvious sufficient condition for Assumption 3.1 and (40) is the following:

C1 (a) There exists $\bar{\alpha} \in (0,1)$ such that for all $(x,a) \in \mathbb{K}$, $\alpha(x,a) < \bar{\alpha}$.

(b) For some constant m, $0 \leq c(x,a) \leq m$ for all $(x,a) \in \mathbb{K}$.

Indeed, under condition C1, $V(\pi,x) \leq m/(1-\bar{\alpha})$ for all $x \in X$ and $\pi \in \Pi$, and $\{w_n\}$ is a bounded sequence which in turn implies (since $w_n \searrow w$ the boundedness of the function w. This fact clearly yields (40).

Other less obvious sufficient conditions are the following (see, e.g., [15, 16, 2]).

C2 (a) Condition C1 (a).

(b) There exist a measurable function $W: X \to (1,\infty)$ and constants $M > 0$, $\beta \in (1,1/\bar{\alpha})$, such that for all $(x,a) \in \mathbb{K}$,

$$\sup_{A(x)} c(x,a) \leq MW(x) \tag{51}$$

and

$$\int_X W(y)Q(dy \mid x,a) \leq \beta W(x). \tag{52}$$

First note that by condition C2 and the Markov property (14), for any policy $\pi \in \Pi$ and initial state $x_0 = x \in X$,

$$E_x^\pi\left[W(x_{n+1}) \mid h_n, a_n\right] = \int_X W(y)Q(dy \mid x_n, a_n) \leq \beta W(x_n), \ \ \forall n \in \mathbb{N}_0. \tag{53}$$

Then, using properties of conditional expectation,

$$E_x^\pi\left[W(x_{n+1})\right] \leq \beta E_x^\pi\left[W(x_n)\right], \ \ \forall n \in \mathbb{N}_0. \tag{54}$$

Iterating inequality (51) we get

$$E_x^\pi\left[W(x_n)\right] \leq \beta^n W(x), \ \ \forall n \in \mathbb{N}_0. \tag{55}$$

Therefore, by condition C2, for any policy $\pi \in \Pi$ and $x \in X$,

$$\begin{aligned} V(\pi,x) &\leq E_x^\pi \sum_{n=0}^{\infty} \bar{\alpha}^n c(x_{n,} a_n) \leq \sum_{n=0}^{\infty} M\bar{\alpha}^n E_x^\pi W(x_n) \\ &\leq \frac{M}{1-\bar{\alpha}\beta} W(x). \end{aligned} \tag{56}$$

Thus, Assumption 3.1 holds.

On the other hand, if $L_+^W(X)$ denotes the subclass of all functions u in $L_+(X)$ such that

$$\|u\|_W := \sup_{x \in X} \frac{h(x)}{W(x)} < \infty, \tag{57}$$

then, because $w_k(\cdot) = V(f_{k+1}, \cdot)$, from (53) and condition C2, we have that $w_k \in L_+^W(X)$ for all $k = 1,2,...$ and

$$\lim_{n \to \infty} E_x^\pi \left[\Gamma_n w_k(x_n) \right] = 0 \quad \forall \pi \in \Pi, x \in X. \tag{58}$$

Since $w \leq w_k$, (40) follows from (55).

Acknowledgements

Work supported partially by Consejo Nacional de Ciencia y Tecnologa (CONACYT) under grant CB2015/254306.

Author details

Fernando Luque-Vásquez and J. Adolfo Minjárez-Sosa*

*Address all correspondence to: aminjare@gauss.mat.uson.mx

Department of Mathematics, University of Sonora, Hermosillo, Sonora, México

References

[1] O. Hernández-Lerma, J.B. Lasserre, Discrete-Time Markov Control Processes: Basic Optimality Criteria. Springer-Verlag, New York, NY, 1996.

[2] O. Hernández-Lerma, J.B. Lasserre, Further Topics on Discrete-Time Markov Control Processes. Springer-Verlag, New York, NY, 1999.

[3] M.L. Puterman, Markov Decision Processes: Discrete Stochastic Dynamic Programming. Wiley, New York, NY, 1994.

[4] Y. Carmon, A. Shwartz, Markov decision processes with exponentially representable discounting. Oper. Res. Lett. 37 (2009), 51–55.

[5] E.A. Feinberg, A. Shwartz, Constrained dynamic programming with two discount factors: applications and an algorithm. IEEE Trans. Autom. Control, 44 (1999), 628–631.

[6] K. Hinderer, Foundations of non-stationary dynamical programming with discrete time parameter, Lecture Notes Oper. Res. 33 Springer, New York, NY, 1970.

[7] M. Schäl, Conditions for optimality in dynamic programming and for the limit of n-stages optimal policies to be optimal, Z. Wahr. Verw. Geb. 32 (1975), 179–196.

[8] Q. Wei, X. Guo, Markov decision processes with state-dependent discount factors and unbounded rewards/costs. Oper. Res. Lett. 39 (2011), 369–374.

[9] J. González-Hernández, R.R. López-Martnez, R. Pérez-Hernández, Markov control processes with randomized discounted cost in Borel space. Math. Meth. Oper. Res., 65 (2007), 27–44.

[10] J. González-Hernández, R.R. López-Martnez, J.A. Minjárez-Sosa, Adaptive policies for stochastic systems under a randomized discounted criterion. Bol. Soc. Mat. Mex., 14 (2008), 149–163.

[11] J. González-Hernández, R.R. López-Martnez, J.A. Minjárez-Sosa, Approximation, estimation and control of stochastic systems under a randomized discounted cost criterion. Kybernetika, 45 (2009), 737–754.

[12] J. González-Hernández, R.R. López-Martnez, J.A. Minjárez-Sosa, J.R. Gabriel-Arguelles, Constrained Markov control processes with randomized discounted cost criteria: occupation measures and extremal points. Risk Decis. Anal., 4 (2013), 163–176.

[13] E.A. Feinberg, P.O. Kasyanov, N.V. Zadoianchuck, Berge's theorem for non compact image sets, J. Math. Anal. Appl. 397 (2013), 255–259.

[14] E.B. Dynkin, A.A. Yushkevich, Controlled Markov Processes. Springer-Verlag, New York, NY, 1979.

[15] E.I. Gordienko, O. Hernández-Lerma, Average cost Markov control processes with weighted norms: existence of canonical policies. Appl. Math. (Warsaw), 23 (1995), 199–218.

[16] E.I. Gordienko, O. Hernández-Lerma, Average cost Markov control processes with weighted norms: value iterations. Appl. Math. (Warsaw), 23 (1995), 219–237.

[17] O. Hernández-Lerma, W. Runggaldier, Monotone approximations for convex stochastic control problems. J. Math. Syst., Est. Control, 4 (1994), 99–140.

Chapter 5

Mathematical Modeling of Isothermal Drying and its Potential Application in the Design of the Industrial Drying Regimes of Clay Products

Miloš Vasić, Zagorka Radojević and Robert Rekecki

Additional information is available at the end of the chapter

http://dx.doi.org/10.5772/64983

Abstract

The processes of simultaneous moisture and heat transfer, which are often nonstationary, and the distinct nature and properties of the material to be dried complicate the description of the drying process. The theory of moisture migration and modeling of drying process has been the subject of many studies. Three theories, the diffusion, the capillary flow, and the evaporation-condensation, have won general recognition for the explanation of moisture transfer in porous media. This study has several objectives. The first one was to present a new method for calculation of the variable effective diffusivity as well as to identify different drying mechanisms and its exact transitions during isothermal drying of clay tiles. The second and main objectives were to analyze all obtained isothermal data, to create a link with the comprehensive theory of moisture migration during drying, and to set up the non-isothermal drying process. The procedure was based on the principle of controlling the mass transport during the drying process. Proposed regimes were consisted from several isothermal segments. Isothermal segments were selected and specificated in accordance with the clay raw material nature and the moisture migration theory.

Keywords: drying regime, effective diffusivity, clay tile, non-isothermal drying, shrinking

1. Introduction

Drying represents very important and complex process in the production of clay tiles, which involves simultaneous heat and mass transfer between the body and the surrounding atmosphere. The whole process consists of several periods characterized with different mechanisms of internal moisture transfer. Until recently, three theories, the diffusion theory, the capillary flow theory, and the "evaporation-condensation theory," have won general recognition for an explanation of moisture transfer in porous media. During drying, several mechanisms are controlling the overall internal moisture transport from the drying material up to its surface. In order to describe the complete internal transport with the same equations as pure diffusion and to take the correction for all secondary types of mass transfer into account, it is suitable to simply replace the pure diffusion coefficient with an effective diffusion coefficient. This procedure was successfully applied in the reference [1]. Within the same reference, it is stated that "the effective moisture diffusivity represents an overall mass transport property of moisture which includes molecular diffusion, the Knudsen diffusion, the non-Fickian or stress-driven diffusion, capillary motions, liquid diffusion through solid pores, vapor diffusion in air-filled pores, vaporization-condensation sequence flow, and hydrodynamic flow mass transfer mechanisms."

Determination of the effective diffusion coefficient is essential for a credible description of the mass transfer process, described by Fick's equation [2]. Description and modeling of drying process based on the calculation of constant effective moisture diffusivity have been the subject of many studies [3–7]. The plot of effective moisture diffusivity vs. time or moisture content (Deff-t or Deff-MR curve) is a good indicator to evaluate and present an overall mass transport property of moisture during isothermal drying. Determination of time-dependent effective moisture diffusivity along with the detection of Deff-MR curves has been reported in several studies [8–11]. The facts that capillary flow is a predominant mechanism within the constant drying period while, in the falling drying period, the evaporation-condensation and vapor diffusion are the predominant mechanisms, have won general recognition for the explanation of moisture transfer in porous media. The comprehensive theory of moisture migration during drying which represents a method useful to trace and quantify all possible mechanisms of moisture transport and their transitions during the isothermal drying process was recently reported [12].

This study has several objectives. The first one was to shortly present the theory of moisture migration along with the method for calculation of the variable effective diffusivity. The next one was to calculate the variable effective diffusivity, to divide drying curve in segments, and to identify all possible mechanisms of moisture transport within a clay roofing tile for several different experiments, in which drying air parameters were constant. The main objectives of this study were to analyze all obtained isothermal data, to create a link with the nature of the raw clay material and the comprehensive theory of moisture migration during drying, and finally to design the non-isothermal industrial drying regime.

2. Methods and materials

After using appropriate initial and boundary conditions, along with reasonable assumptions, Cranck has presented the analytical solution of Fick's second diffusion law for several standard

http://dx.doi.org/10.5772/64983

geometries, such as tile, cylinder, and sphere [13]. Hence, on the base of the lumping approximation, which assumes that the effective diffusivity is an overall mass transport property, the Cranck solution for tile geometry can be expressed as Eq. (1):

$$MR = \frac{X - X_{eq}}{X_0 - X_{eq}} = \frac{8}{\pi^2} \sum_{n=1}^{\infty} \frac{1}{(2n+1)^2} exp\left(\frac{-(2n+1)^2}{4} \right) \pi^2 \frac{D_{eff} t}{l^2} \quad (1)$$

In order to estimate effective diffusivity from Eq. (1), many researchers were using a simplifying method [14, 15]. They assumed that for sufficiently long drying times, the terms in the infinite summation series in Eq. (1) converge rapidly and in most cases may be accurately approximated by the first term of the series. This method is commonly known as "simplified slope" model. Two programs for determination of effective diffusion coefficient, based on mathematical calculation of Fick's second law and the Cranck diffusion equation, were recently presented for clay tiles [2, 6, 7].

2.1. Method for estimation of the time-dependent effective diffusivity

Zagrouba was one of the first researchers that had reported the "slope" method as a possible solution for estimation of the dependent effective diffusivity at various moisture contents for clay materials [11]. The Fourier diffusion number (F_0) has to be calculated from Eq. (2), while constant diffusion coefficient D_0 is obtained from the "simplified slope" method. The dependent effective diffusivity D_{eff} is calculated from Eq. (3). This method has been widely applied in several studies such as [8, 9, 16]:

$$F_0 = \frac{D_0 t}{l_0^2} \quad (2)$$

$$D_{eff} = \frac{\left(\frac{\partial MR}{\partial t} \right)_{exp}}{\left(\frac{\partial MR}{\partial F_0} \right)_{th}} l_0^2 \quad (3)$$

2.1.1. Modified slope method

Better and more accurate solution, called "modified slope" method, for calculation and estimation of time-dependent effective diffusivity based on the "slope" method was presented [12] and recently applied in two studies [17, 18]. The essence of this new model is reflected in replacing the constant term l_0 in Eqs. (2) and (3) with an interchangeable $l_{(t)}$ term and by using the model that includes shrinkage, which is presented in the study [7], for calculation of the constant diffusion coefficient D_0. Interchangeable $l_{(t)}$ term must be registered experimentally. The final calculation formula is presented in the form of Eq. (4):

$$D_{eff_{(t)}} = \frac{\left(\frac{\partial MR}{\partial t}\right)_{exp}}{\left(\frac{\partial MR}{\partial F_0}\right)_{th}} l_{(t)}^2 \tag{4}$$

2.1.2. Moisture migration theory

Typical curves, which represent the dependence of the effective moisture diffusivity as a function of the moisture content or drying time obtained using calculation method presented in the study [12], are presented in **Figure 1**.

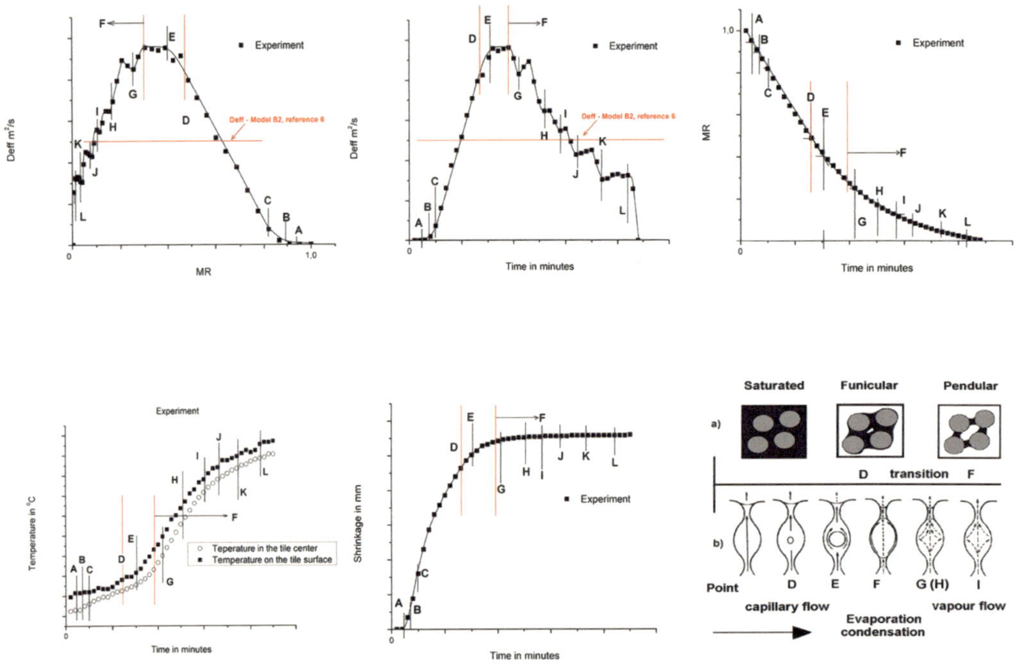

Figure 1. Typical time-dependent effective moisture diffusivity curves (Deff-MR and Deff-t).

It is important to highlight the significance of the characteristic pattern shown in **Figure 1**. It indicates all possible mechanisms of moisture transport along with their transitions from one to another during the constant and the falling drying period for isothermal experiments.

At the beginning of the drying process, effective moisture diffusivity values are equal to zero until characteristic point A is reached. This drying period is commonly known as the "initial heating segment." It is a relatively short period. The quantity of evaporated water is small, and shrinkage of the green body (clay tile) is not detected. The clay tile surface is heating up from its starting temperature up to the wet-bulb temperature. Moisture diffusivity values are practically zero suggesting that the overall mass transport is negligible.

After the initial heating period is over, the effective moisture diffusivity values are increasing as the moisture ratio is decreased, until characteristic point E is reached. This period is commonly known as the "constant drying rate period." Throughout this period, the surface of the green body is constantly covered with a continuous film of water. The surface temperature is constant and has a value corresponding to the wet-bulb temperature of the air. The shrinkage of the green body is characteristic for this drying period. The end of the constant rate period is marked by a maximum in capillary pressure. Cracking of the green body is most likely to occur at this point of the drying process [19–21].

From point A to point B, liquid is transporting through the biggest capillaries. This transport is caused by a gradient of the capillary potential and is commonly known as the "capillary pumping flow." Throughout this period, the quantity of evaporated water and detected sample shrinkage is relatively small. The "hydrodynamic flow" (caused by viscosity) is negligible and that the overall mass transport is governed only by capillary pumping flow mechanisms.

From point B up to point D, liquid is transported simultaneously by two mechanisms. The main mechanism is capillary pumping flow, caused by the gradient of the capillary potential originating from still saturated capillaries, and the second one is hydrodynamic liquid flow in pores, which arises from the difference in total pressure caused by friction. Capillary pumping flow up to point C was caused by the still present macro capillaries, while the same flow up to point D was caused by the presence of mezzo capillaries. Throughout this period, the quantity of evaporated water and detected sample shrinkage is considerably higher than in the already presented drying segments. Point D is commonly known as the "upper critical point" which is indicating the beginning of the transition from the "funicular" to the "pendular" state. It is important to highlight that in the "funicular" state, continuous threads of moisture are present in the pores, while in the "pendular" state this is not the case. From this point on, the clay tile surface is not fully covered by a water film. "Dry" patches will appear on the surface for the first time. The drying front is starting to recede.

From point D up to point E, liquid and vapor are transported simultaneously. Liquid is transported by three mechanisms. The main one is capillary pumping flow. Capillaries in the funicular state are generating the pressure gradient which secures the capillary pumping flow. The difference in total pressure caused by friction provides hydrodynamic liquid flow in pores which represents the second transport mechanisms. The concentration gradient of the liquid in the pores is the driving force for the liquid diffusion which represents the third transport mechanisms. The difference in total pressure caused by friction secures the hydrodynamic vapor flow in pores. Deviation from a constant drying rate is first registered as point E, which is commonly named the "critical" point. A partially wet surface is able to provide a constant or a falling drying period depending on the fraction of wet surface and the boundary layer thickness. The influence of a partially wet surface on the transition from a constant rate to a falling rate of drying is described in the study [22].

From point E up to point F, the fraction of the wet surface is decreased until the "last" wet patches disappeared from the surface. This point is commonly known as the "lower" critical point which indicates the end of the transition from the "funicular" to the "pendular" state. In other words, the moisture content is decreased, and the gas bubbles attained the dimensions of the pores, breaking the continuous threads of moisture in the pores. Moisture is transported up to the surface by creeping along the capillary when the liquid is in the funicular state or by the successive evaporation-condensation mechanism between liquid bridges. When the "pendular" state is reached, there is no further contraction of the drying body, and consequently the possibility of the drying body to crack is extremely small.

During the drying process, the temperature increase is registered. The temperature is increasing slowly from point D up to point E, and then moderate temperature increase is registered up to point F. After point F is reached, the temperature of the system is rising rapidly up to point I. From point I up to point K, the temperature is rising very slow and is practically close to the level off at the so-called pseudo-wet-bulb temperature. The temperature is increasing again just before the end of the drying process until it reached the level of the final temperature.

With further drying from point F to point G, the moisture is transported up to the surface by the successive evaporation-condensation mechanism between liquid bridges. Simultaneously with the evaporation and condensation mechanisms, liquid starts to evaporate within the pore space at a growing rate, causing the vapor pressure to increase. These two mechanisms move the "pendular" water up to the surface.

The G–H segment is consisted from two parts. Locally produced vapor is accumulating within the pores causing a local increase in the effective diffusivity within the first G–H segment. In one moment, the local vapor pressure is exceeding the critical value, and the vapor is practically "blown" away up to the surface. During the vapor release, some liquid bridges of "pendular" water are also transported. This kind of moisture transport is accompanied with the local decrease of the effective diffusivity value which is characteristic for the second G–H segment. The movement of the remaining "pendular" water in the H–I segment is mostly caused by the vapor pressure existing within the pore space.

After "pendular" water is removed, the evaporation occurs only inside the body, and the temperature of the surface approaches the so-called pseudo-wet-bulb temperature, and at the end of the drying, it reaches ambient temperature. This I–L segment is commonly known as the "diffusion period." It is divided into three parts. The first one "I–J" represents pure molecular diffusion, while the second "J–K" and third "K–L" ones represent, respectively, transitional and the Knudsen diffusion.

2.2. Experimental

The raw material, used in this study, was obtained from the largest roofing tile manufacturer in Serbia. "Kanjiza's" clay raw deposit is formed from two layers. The first layer is commonly known as the "yellow clay." It contains a relatively small amount of clay minerals (under 23 wt.%), but it is rich in quartz, carbonates (above 20 wt.%), and feldspar minerals. The second layer commonly known as the "blue clay" predominantly contains clay minerals: illite, smectite, chlorite, and kaolinite. The industrial raw material mixture and the one used in this study consist, respectively, 80 and 20 wt.% of blue and yellow clay [23].

Initial characterization of Initial characterization of the raw material has included determination of particle size distribution (PSD) analysis, standard silicate chemical analysis (SSA), qualitative and semiquantitative XRD analysis, and TGA (TG analysis). Standardized procedures, described, respectively, in SRPS U.B1.018:2005 and SRPS B.D8.210:1982 norms, were used for PSD and SSA determination. Qualitative and semiquantitative XRD analysis and TGA (TG analysis) were reported in the study [24].

After initial characterization, the raw material was homogenized and prepared for the forming process. During the homogenization process, the raw material was moisturized and milled using laboratory differential mills.

Laboratory roofing tile samples 120 × 50 × 14 mm were formed in a laboratory extruder "Hendle" type 4, under a vacuum of 0.8 bar. Formed samples were packed into plastic bags which were afterward sealed and put into a glass container with lid. Glass containers with samples were kept in the air-conditioned room in which temperature and relative humidity were maintained, respectively, at 25°C and 65%. This procedure allows the minimal moisture content fluctuations within the stored samples.

Series of isothermal drying curves was recorded. Laboratory recirculation dryer in which drying parameters (humidity, temperature, and velocity) could be programmed, controlled, and monitored was used. Regulation of wet air parameters within the range of 0–125°C, 20–100%, and 0–3.5 m/s with accuracies of ±0.2°C, ±0.2%, and ±0.1% for temperature, humidity, and velocity, respectively, was limited by the laboratory dryer design. The mass of the samples and their linear shrinkage were continually monitored and recorded during the experiments. The accuracies of these measurements were 0.01 g and 0.2 mm, respectively. Experimental conditions presented in **Table 1** were used in the present study. Each experiment was repeated two times.

Experiment	Group	Air humidity, *V*/%	Air temperature, *T*/°C
1	I	80	35
2	I	80	40
3	II	70	40
4	II	70	45
5	III	60	40
6	III	60	45

Table 1. Experimental conditions.

The modified slope method was used to calculate the functional dependence of the effective diffusivity vs. moisture content (Deff-MR), to divide obtained curves in segments, and to identify all possible mechanisms of moisture transport within each drying segment. Roofing tile samples were afterward dried to constant mass. Dried samples were heated in oxygen atmosphere with the heating rate of 1.4°C/min from room temperature up to 610°C and further with the heating rate of 2.5°C/min up to the 1000°C. Samples were kept at 1000°C for 2 hours. Flexural strength was determined on dried (DSFS) and fired samples (FSFS) using the procedure described in EN 538 norm.

Obtained data were analyzed and used to set up several non-isothermal drying regimes. Drying air parameters which were maintained in each proposed drying regime are presented in **Table 2**. Duration of the approximately isothermal drying segments was detected from the isothermal curves Deff-MR. Tiles were then dried to constant mass and fired using the same heating rate as one previously mentioned. DSFS and FSFS were determined. Twist coefficient (TWC) and longitudinal camber (LOC) coefficient and transverse camber (TRC) coefficient were determined on fired samples using the procedure described in EN 1024 norm.

Experiment	**Segment**				
	I	**II**	**III**	**IV**	**V**
7	Exp. 1	Exp. 2	Exp. 3	Exp. 4	70°C/40%
8	Exp. 2	Exp. 3	Exp. 5	Exp. 6	70°C/40%

Table 2. Experimental conditions—proposed drying regimes.

3. Results and discussion

Results of several analyses, used for initial characterization of the raw material, are presented in **Table 3**. The mass content of SiO_2, Al_2O_3, CaO, and MgO obtained by SSA analyses is indicating the presence of free quartz, feldspars, clay minerals, and carbonate minerals in the analyzed raw material. Results of qualitative mineralogical and XRD analyses, reported in the study [24], have confirmed the presence of quartz, feldspars (orthoclase), illite, muscovite, kaolinite, montmorillonite, chlorite, calcite, and dolomite in the analyzed raw material. Semiquantitative XRD mineralogical analysis has quantified the presence of previously mentioned minerals. The mass content of clay, silt, and sand obtained by PSD analyses is indicating that the analyzed raw material is classified as clay loam suitable for clay roofing tile production.

All possible mechanisms of moisture transport and their transition from one to another during drying for isothermal experiments are identified and are shown in **Figure 2**.

http://dx.doi.org/10.5772/64983

Chemical composition in wt%		Mineralogical composition in wt%	
Loss on ignition 1000°C	6.24	Quartz	29
SiO_2	55.30	Feldspars (orthoclase)	17
Al_2O_3	16.00	Illite and muscovite	16
Fe_2O_3	6.12	Kaolinite	6
CaO	7.50	Montmorillonite	5
MgO	3.59	Chlorite	6
SO_3	0.00	Calcite	7
S^{2-}	0.00	Dolomite	6
Na_2O	1.10	Particle size composition in wt%	
K_2O	3.52	Clay (<2 μm)	46.2
MnO	0.08	Silt (2–20 μm)	29
TiO_2	0.54	Sand (>20 μ m)	24.2

Table 3. Initial characterization of the raw material.

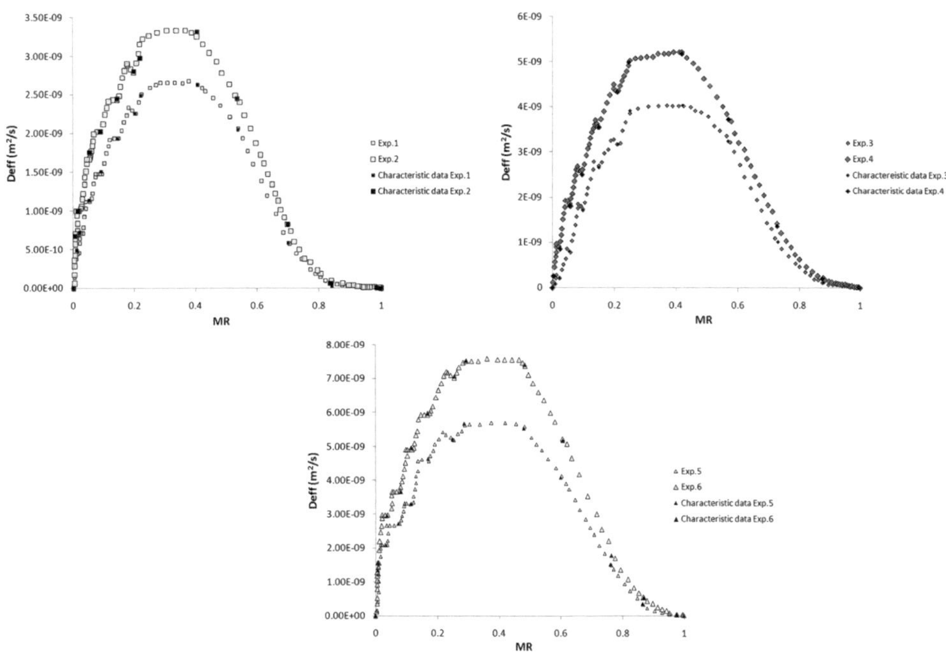

Figure 2. Estimated Deff-MR curves for isothermal experiments.

The procedure for setting up the non-isothermal drying regime that is consistent with the theory of moisture migration during drying was based on the principle of controlling the mass transport during the drying process and has demanded to divide the drying process into five segments. In each of these segments, approximately isothermal drying conditions were maintained (see **Table 2**).

The main functions of the first segment are to restrain the moisture transport (evaporation), through the boundary layer between material surface and the bulk air, and to heat the ceramic body to the temperature of the drying air. That is the reason why high values of the drying air humidity in the first segment were selected. In order to fulfill previously mentioned requirements, this drying segment is over when characteristic point C is reached. During the second drying segment, external transport (surface evaporation) and internal transport (of liquid water from the ceramic body up to the surface) have to be increased and simultaneously harmonized in such a way that the drying surface remains fully covered by a water film. That is the reason why in most cases drying air humidity in this segment is reduced. Its absolute value is still relatively high. This will increase the evaporation driving force and consequently will speed up the drying process. Drying air temperature in this segment may slightly increase compared to the previous segment. This will moderately increase the capillary transport as well as the drying rate. The second segment starts and ends when characteristic points C and D ("upper critical point") are, respectively, reached.

The third and fourth segment represents together a transitional drying period in which the sample is gradually shifting from "funicular" to "pendular" state. A higher fraction of wet surface and a thicker boundary layer produce and favor a constant drying rate period, while a smaller fraction of the surface and a thinner boundary layer shell favor and produce a falling drying rate. The main function of the third segment is to provide the conditions that will lead to the fact that partially wet surfaces provide a constant rate of drying. That is the reason why the humidity and temperature of the drying air within the third segment have to be carefully selected. Further reduction of the drying air humidity (see **Table 2**) will increase the evaporation driving force (external surface evaporation) and consequently will speed up the drying process. Drying air temperature in this segment may slightly increase compared to the previous segment. This will increase the capillary transport as well as the drying rate. The third segment starts and ends when characteristic points D and E are, respectively, reached.

Within the fourth segment, the fraction of the wet surface decreased until the "last" wet patches disappeared from the surface. At the end of the fourth segment, the system had reached the "pendular" state, and there is no further contraction of the drying body. The main function of this segment is to simultaneously harmonize the liquid transport originating from the pores which are near or just below the "dry" patches on the surface and are still in the funicular state with the liquid flow originating from the surface "wet" patches. That is the reason why drying air humidity in this segment is not reduced (see **Table 2**).

Further increase of the drying air temperature has a positive influence which has led to the liquid transport enhancement. The fourth segment starts and ends when characteristic points E and F ("lower critical point") are, respectively, reached. The main function of the fifth segment is to maximally facilitate the internal moisture transport up to the surface. That is the reason why drying air humidity in the fifth segment is further reduced, while drying air temperature is maximally increased [25].

Characteristic data for isothermal experiments from point A up to point F are presented in **Table 4**.

Exp.		1	2	3	4	5	6
Time (min)	A	58	51	49	36	23	21
	B	161	148	109	89	49	45
	C	280	253	190	170	82	75
	D	424	387	284	268	145	133
	E	526	484	360	338	187	170
	F	769	700	518	491	291	268
	TDT	1420	1250	1190	1050	730	625
	TDT—total drying time						

Table 4. Characteristic data for isothermal experiments from point A up to point F.

Duration of the approximately isothermal drying segments was not specified by experience or by trial-and-error method. It was detected from the appropriate isothermal Deff–MR curves (see **Figure 1** and **Table 4**). General procedure will be explained on experiment 7. Duration of the first segment was the same as the duration of the drying process in the case of experiment 1 from the beginning up to the characteristic point C. Duration of the second segment was the same as the duration of the drying process in the case of experiment 2 from the characteristic point C up to the characteristic point D. Duration of the third segment was the same as the duration of the drying process in the case of experiment 3 from the characteristic point D up to the characteristic point E. Duration of the fourth segment was the same as the duration of the drying process in the case of experiment 4 from the characteristic point E up to the characteristic point F. Duration of the fifth segment was limited to 90 minutes. This procedure was used to specify the duration of drying segments in each proposed drying regime. Calculated results are presented in **Table 5**.

Drying segment	Segment duration t (min)	
	Exp. 7	Exp. 8
1	280	253
2	134	94
3	76	42
4	153	98
5	90	90
TDT	733	577

Table 5. Calculated segment duration within proposed drying regimes.

It is important to define the minimum requirements for dried clay roofing tile which if satisfactory will ensure that the product is able to perform its function. In other words, dried clay roofing tiles has to be dried without cracks. Minimal flexural strength of dried and fired samples has to be, respectively, at least 0.73 and 1.2 kN (see EN 1304 norm). The mean value of the twist coefficient (TWC) and the mean value of the longitudinal camber (LOC) and transverse camber (TRC) coefficients calculated as described in EN 1024 norm shall comply, respectively, with the requirements stated in **Tables 1–3** presented within the EN 1304 norm. Proposed drying regimes were tested. Clay roofing tiles were dried without cracks. Flexural strength of dried and fired clay tiles (DSFS and FSFS) are presented in **Table 6**.

Experiment	DSFS (kN)	1000^{0}C
		FSFS (kN)
1	0.99	2.73
2	0.97	2.65
3	0.92	2.67
4	0.82	2.65
5	0.80	2.12
6	0.75	2.08
7	0.93	2.65
8	0.81	2.42

Table 6. Mechanical properties of dried and fired samples.

Experiment	Twist coefficient	Cambers R (%)	
	C (%)	Longitudinal	Longitudinal
1	0.29	0.33	0.33
2	0.45	0.50	0.50
3	0.90	0.68	0.68
4	1.23	0.87	0.87
5	0.72	0.79	0.79
6	1.18	1.05	1.05
7	0.48	0.51	0.51
8	0.71	0.69	0.69

Table 7. Twist and camber coefficients.

The mean TWC value and the mean LOC and TRC values for isothermal and non-isothermal drying regimes are presented in **Table 7**. Detailed analysis of data presented in **Table 7** has revealed that mean TC, RL, and RT is increasing with the increase of the drying air temperature as well as that under the same drying air temperature, TWC, LOC, and TRC values, presented in different experimental groups, is also increasing with the decrease of air relative humidity (see **Tables 1** and **7**). Maximally allowed deviation for TWC, LOC, and TRC is 2%. This criterion is defined in **Tables 1–3** of the EN 1304 standard.

TWC, LOC, and TRC values can be used as a good indirect indicator of the stress generation. In other words, higher TWC, LOC, and TRC values are correlated with the higher stress generation. The drying air with higher temperature and lower relative humidity leads to more rapid generation of the stress in the samples during drying which will result with the lower shape regularity (higher coefficients) and lower mechanical properties of the dried samples.

The shortest total drying time (TDT) for isothermal and non-isothermal drying regimes was, respectively, registered in experiments 8 and 6 (see **Tables 4** and **5**). The difference between TDT values related to previously mentioned experiments is relatively small. The lowest DSFS and FSFS values along with the highest mean TWC, LOC, and TRC values for non-isothermal drying regimes were registered in experiment 8. It is important to point out that dried and fired clay roofing tiles in each proposed non-isothermal experiment have satisfied previously mentioned flexural strength as well as shape regularity (twist and cambers) criteria. That is the reason why in this study the lowest TDT value was used as a final criterion for selection of the experiment 8 drying regime as optimal (see **Table 5**, experiment 8).

4. Conclusion

The procedure for setting up the non-isothermal drying regime that is consistent with the theory of moisture migration during drying has demanded to divide the drying process into five segments. For the first time, the choice of isothermal segments specification was achieved in accordance with the theory of moisture migration during drying and with the nature and the properties of the clay raw materials. Duration of the approximately isothermal drying segments was not specified by experience or by trial-and-error method. It was detected from the appropriate isothermal Deff-MR curves. Proposed drying regimes were tested. Dried clay roofing tiles have satisfied all requirements related to the shape regularity and mechanical properties as defined in EN 1304 norm. Finally, experiment 8 was chosen as the optimal drying regime. Semi-industrial trials have shown that the proposed drying regimes obtained from Deff-MR curves can be implemented in real industry system. Namely, the design of the optimal drying curve along with the drying time reduction and higher utilization of the dryer is possible without the fear of generating higher scrap rate. The next step is to apply the presented procedure and to find a way to distinguish the influence of shape factor, forming history and drying parameters on the quality of the dried tiles.

Acknowledgements

This paper was realized under the project ИИИ 45008 which was financed by ministry of education and science of Serbia as well as the company "Potisje Kanjiža".

Author details

Miloš Vasić[1*], Zagorka Radojević[1] and Robert Rekecki[2]

*Address all correspondence to: milos.vasic@institutims.rs

1 Institute for the Testing of Materials, Belgrade, Serbia

2 Potisije Kanjiža, Kanjiža, Serbia

References

[1] Efremov G., Kudra T. Calculation of the effective diffusion coefficients by applying a quasi-stationary equation for drying kinetic. Drying Technology. 2004;22(10):2273–2279.

[2] Vasić M., Radojević Z., Grbavčić Ž. Methods of determination for effective diffusion coefficient during convective drying of clay products. In: Valaškova M., Martynkova S.G., editors. Clay Minerals in nature—their characterization, modification and application. Croatia: Intech; 2012. p. 295–313.

[3] Vasić M., Radojević Z. Calculation of the effective diffusion coefficient. International Journal of Modern Manufacturing Technologies. 2011;3(1):93–98.

[4] Vasić M., Radojević Z., Grbavčić Ž. Calculation of the effective diffusion coefficient during the drying of clay samples. Journal of the Serbian Chemical Society. 2012;77(4): 523–533.

[5] Waccarezza L.M., Lombardi J.L., Chirifle J. Effects on drying rate of food dehydration. Canadian Journal of Chemical Engineering. 1974;52(5):576–579.

[6] Vasić M., Radojević Z., Grbavčić Ž. Determination of the effective diffusion coefficient. Romanian Journal of Material. 2011;42(1):169–176.

[7] Vasić M., Grbavčić Ž., Radojević Z. Determination of the moisture diffusivity coefficient and mathematical modeling of drying. Chemical Engineering and Processing. 2014;76:33–44.

[8] Pinto L.A.A., Tobinaga S. Diffusive model with shrinkage in the thin-layer drying of fish. Drying Technology. 2006;24(4):509–516.

[9] Lopez I.I.R., Espinoza H.R., Lozada P.A., Alvarado A.M.G. Analytical model for moisture diffusivity estimation and drying simulation of shrinkable food products. Journal of Food Engineering. 2012;108:427–435.

[10] Batista M.L., Cezar A.R., Pinto L.A.A. Diffusive model with variable effective diffusivity considering shrinkage in thin layer drying of chitosan. Journal of Food Engineering. 2007;81(1):127–132.

[11] Zagrouba F., Chemki S. Water diffusion coefficient in clay material from drying data. Desalination. 2005;185:491–498.

[12] Vasić M., Grbavčić Ž., Radojević Z. Analysis of moisture transfer during the drying of clay tiles with particular reference to an estimation of the time-dependent effective diffusivity. Drying Technology. 2014;32(7):829–840.

[13] Cranck J. The Mathematics of Diffusion. New York, USA: Oxford University Press; 1975.

[14] Aghbashlo M., Kianmehr M.H., Akhijahani H.S. Influence of drying conditions on the effective moisture diffusivity, energy of activation and energy consumption during the thin-layer drying of berberis fruit (Berberidaceae). Energy Conversion and Management. 2008;49:2865–2871.

[15] Doymaz I. Convective drying kinetics of strawberry. Chemical Engineering and Processing. 2008;47:914–919.

[16] Azzouz S., Guizani A., Jomaa W., Belghith A. Moisture diffusivity and drying kinetic equation of convective drying of grapes. Journal of Food Engineering. 2002;55:323–330.

[17] Vasić M., Radojević Z. Non-isothermal drying process optimisation—drying of clay tiles. In: 3th International Conference Modern Technologies in Industrial Engineering (ModTech2015); 17–20.06.2015; Mamaia, Romania. England: IOP Conf. Series: Materials Science and Engineering 95 (2015) 012025; 2015. p. 1–7. DOI: 10.1088/1757-899X/95/1/012025.

[18] Vasić M., Radojević Z. Setting up the drying regimes based on the theory of moisture migration during drying. In: Dumitru Nedelku, editor. Modern Technologies in Industrial Engineering (ModTech 2016), Book of Abstracts; 15–18.06.2016; Iasi, Romania. Iasi, Romania: Modtech Publishing House; 2016. p. 116.

[19] Kowalski S.J. Thermomechanical approach to shrinking and cracking phenomena in drying. Drying Technology. 2001;19(5):731–765.

[20] Kowalski S. J., Banaszak J. Modeling and Experimental Identification of Cracks in Porous Materials during drying. Drying Technology. 2013;31(12):1388–1399.

[21] Aresnović M., Radojević Z., Stanković S., Lalić Ž., Pezo L.L. What to expect from heavy clay? Ceramic International. 2013;39(2):1667–1675.

[22] Shiming Y. Z., Dengying L. Mechanisms and mathematical model of heat and mass transfer during convective drying of porous materials. Heat Transfer—Asian Research. 1999;28(5):337–351.

[23] Rekecki R., Kuzmann E., Homonnay Z., Ranogajec J. Mossbauer and X-ray study of the firing process for production of improved roofing tiles. Hyperfine Interact. 2013;217(35):27–35.

[24] Rekecki R., Ranogajec J. Design of ceramic microstructures based on waste materials. Processing and Application of Ceramics. 2008;2(2):89–95.

[25] Lalić Ž., Aresenović M., Janaćković Đ., Vasić M., Radojević Z. Influence of increased temperature on clay fast drying process. Romanian Journal of Materials. 2009;39(3): 175–179.

Chapter 6

Financial Feasibility Analysis of Natura Rab Business: Case Study

Karmen Pažek, Matija Kaštelan, Martina Bavec,
Črtomir Rozman and Jernej Prišenk

Additional information is available at the end of the chapter

http://dx.doi.org/10.5772/65307

Abstract

In 2015 Natura Rab decided to provide three very important investments that will greatly change and facilitate its future business activities, especially the first project. The first and largest financial investment is the construction of the new organic shop with products at the central farm called Natura Rab. The second investment project is the new 2500 m^2 olive plantation. The third investment in the analyzed family company is related to the beekeeping sector, and it involves several activities like buying new beekeeping equipment and new work vehicle. Before implementing the three investment projects, some financial parameters for the further assessment of investments were used, such as the net present value (NPV) and the internal rate of return (IRR). The investment value of the new shop is 38315.88 €, and the annual cash flow is 13,288 €. The net present value at the discount rate of 5.5% in the fourth year is 8260.55 €. The internal rate of return is 14.51%. The investment value for the second project, the new olive plantation, is 6620 €, and the annual cash flow is 2664.02 €. The net present value at the discount rate of 5.5% in the third year is 567.35 €. The internal rate of return is 10.04%. The investment value of the beekeeping sector for this year is 18428.50 €, and the annual cash flow is 41537.20 €. The net present value at the discount rate of 5.5% after the first year is 20943.25 €.

Keywords: Business project, Financial feasibility, Organic farming, Investments, CBA

1. Introduction

The aim of a successful agriculture business in the twenty-first century is to achieve high profit in the shortest time possible, regardless of the type of agriculture production. This kind of business activity requires constant care and concentration but also the monitoring of competition. The success or failure of agricultural production is directly connected with the competence of a farmer, or, in other words, the one with a higher profit in agricultural production has managed to optimize investment costs in agricultural production [1].

In the field of agricultural economics, there are two main types of costs. The permanent (fixed) costs that are substantially independent of the production volume and variable costs that are significantly altered in changes in production output. It is the basic classification of costs. However, the dilemma lies in something else. Functionality of brought investment will be revealed with more precise data than just fixed and variable costs, which have direct or indirect impact on the investment [1]. So in financial analysis, there is a wide variety of different costs. In [2] author explains six most important management views in agricultural projects (cited in [3]):

—*Technical aspects* (analysis of the availability of production means, determining the quantity of inputs and associated production levels, identification of the existing relations between different business entities of agricultural production, etc.).

—*Institutional-organizational aspects* (the study of the institutional environment, within which the performance of a given investment project is predicted, especially in light of the evaluation of its organizational tidiness).

—*Commercial aspects* (dynamic evaluation of different production options while checking the abilities in marketing of agricultural products).

—*Financial aspects* (financial evaluation of the meaningfulness of investment in agriculture by identifying the resulting income or loss).

—*Economic aspects* (evaluation of the real contribution of specific agricultural project to positive growth of the entire economy).

—*Socio-social aspects* (specific socioeconomic analysis of the effects that can go in national economic system by individual investment project in agriculture).

The organic farm that was analyzed wanted to invest in three different projects in 2015 in order to improve sales and production conditions on the farm. All this progress goes for increasing productivity of the family company. Our own considerations and head calculations for projects were not enough and not correct. Economic parameters were used for the assessment of investments. A dynamic method of investment evaluation (CBA analysis) was applied to all three projects. The main goal was to develop a model for the assessment of the investment. It was examined by net present value (NPV) and internal rate of return (IRR). Planned annual cash flow and calculations for products were also developed with all additional costs in order to rate the investment more precisely, which gave us a realistic picture of the exact return of each investment. Economically efficient agricultural media system is the fundamental goal of the guidelines of every agricultural policy. Assessment of investment projects in agriculture is a multifaceted process [3]. The evaluation of specific agricultural projects is planned like evaluation of certain investments and has to be based on a variety of input costs and bring some types of benefits. In agricultural economics, in this case we are talking about the Cost-benefit analysis (CBA). That is a comparative analysis of the total cost and total revenue of the agricultural project.

2. Materials and methods

On 4000 m^2 of land in the Barbat village at the south end of the island of Rab, a unique organic farm Natura Rab is developed. It is a family-run business comprising of both production and sale, growing typical medicinal and cultivated herbs, beekeeping, and sale of our own products right at our front door.

- In the research we will present three different projects we invested in this year. Project 1 is the new organic shop, project 2 is the new olive plantation, and project 3 is investment in beekeeping facilities. Further in text each project is presented separately, the reasons for their implementation are given as well as our ultimate goals. For each project identical tables in Microsoft Word and Microsoft Excel were made. All the necessary parameters are provided in order to have precise results and correct economic assessment of investment. Fixed and variable costs were separated in the CBA analysis. Costs of annual production of certain final products are also considered which serve as the production output. For each project there is a product table with explanations. It was necessary to be as realistic as possible to avoid imaginary situations, because then we lie to ourselves and the return of investment is not authentic. Technical specifications are made in detail especially in the first two projects that were technically demanding and the work dynamics was longer. Using the program developed in Microsoft Excel, it is easier to calculate the net present value and internal rate of return for each project using input data like amount of investment, annual cash flow, and discount rate with fixed value of 5.5% in all three projects.

3. Methodology of total cost calculation

3.1. Costs

Before we start describing complex economic parameters in this context, it is necessary to explain the theory of production costs. Costs are an integral part of each production process, and they appear as a result of different activities in the production chain [1]. We distinguish between fixed and variable costs, due to the fact that there are some costs that change during longer time period.

Fixed costs, which are independent of the production volume, reflect the use of fixed production assets. For example, fixed costs are land rent, interest related to the acquisition of agricultural land, various mortgage, and insurance premiums. However, various types of amortizations in agriculture production (buildings, machinery) are defined also like fixed costs which relate to noncash payments in agriculture [1].

Variable costs are dependent on the volume of production. This group of costs represents a wide range of various agricultural inputs and costs related to their use (pesticides, fertilizers, seeds, animal feed) [1].

The total cost of production as the sum of all production costs

$$TC = FC + VC \tag{1}$$

FC, fixed costs (€); VC, variable costs (€); and TC, total costs (€).

3.2. Investment costs

Investment costs are present in all three business projects. Consequently, they indicate the amount of each investment. In particular, they are separated in the first two projects, like investment construction costs and investment material costs.

3.3. CBA analysis in agricultural projects

The CBA analysis is the main methodological tool in the process of evaluation of specific agricultural projects or investments made in farming industry or some other agriculture types [3]. Comparative analysis of total revenues and total costs provides an answer to the question of selection of some investment projects in agriculture. All potential costs and revenues must be identified. As we look at all costs and revenues, we have to decide which investment projects will be selected and which will be denied [3]. The most important are the net present value (NPV) and the internal rate of return (IRR).

3.4. Annual cash flow (FR)

The annual cash flow is calculated as the difference between the total revenues and total costs:

$$FR = TR - TC \tag{2}$$

FR, annual cash flow (€); TR, total revenues (€); and TC, total costs (€).

3.5. Method of net present value (NPV)

In financial terms, the net present value (NPV) is defined as the sum of the present values (PVs) of incoming and outgoing cash flows over a period of time. Incoming and outgoing cash flows can also be described as benefit and cost cash flows, respectively [4]. It is a basic norm for financial decision-making. NPV encompasses the concept of the time value of money taking into account the present and future value of money such as in times of inflation [5]. Net present value (NPV) is determined by calculating the costs for each period of an investment, and after the cash flow is calculated, the present value (PV) of each period is achieved by discounting its future value at a periodic rate of return [6]. NPV is the sum of all the discounted future cash flows. NPV is a useful tool to determine whether a project or investment will result in a net profit or a loss. A positive NPV results in profit, while a negative NPV results in loss (**Table 1**) [4]:

If...	It means...	Then...
NPV > 0	The investment would add value to the firm	The project may be accepted
NPV < 0	The investment would subtract value from the firm	The project may be rejected
NPV = 0	The investment would neither gain nor lose value for the firm	We are not sure whether to accept or reject the project. This project adds no monetary value. Decision should be based on other criteria, e.g., strategic positioning or other factors not explicitly included in the calculation

Table 1. NPV use in decision-making process.

$$\mathrm{NPV} = -I + \sum_{i=1}^{n} \frac{\mathrm{TR} - \mathrm{TC}}{(1+r)^{t}} \tag{3}$$

NPV, net present value (€); I, amount of each agricultural investment (€); TR, total revenue (€); TC, total costs (€); r, average annual discount rate (%); and t, time period (number of years).

3.6. Method of internal rate of return (IRR)

Internal rate of return is the second important decision-making tool. Associated concept of net present value is internal rate of return, which is not served by a nominal value but the percent (interest) on the basis, and it still financially justifies the implementation of a certain investment in agriculture [3]. Simply put, the internal rate of return is a rate where NPV of the project is equal to zero that can be seen in the formula below:

$$\mathrm{IRR} = -I + \sum_{i=1}^{n} \frac{\mathrm{TR} - \mathrm{TC}}{(1+r)^{t}} = 0 \tag{4}$$

IRR, internal rate of return (%); I, amount of each agricultural investment (€); TR, total revenue (€); TC, total costs (€); r, average annual discount rate (%); and t, time period (number of years).

4. Results and discussion

4.1. Project 1: investment in the new organic shop

In 2015, we decided to expand our existing organic shop. We want to have a unique space which is larger and more comfortable than the previous one, to obtain space for tasting and sale of our organic products. The new organic shop becomes our only direct sales channel on the island, because we closed the second shop in the old town of Rab. Through the years we became a must-see station of our island of Rab. With good organization, development, and quality, we became a well-established company, and our old customers always keep coming back to buy and enjoy our products, and there always new ones as well. We create long-term relationships with customers based on high development of trust, as one of our main business goals.

In our new organic shop, we want to show all the riches of our hundred-year-old family tradition by representing old agricultural tools and beekeeping equipment of our ancestors; things that were hidden for years in cellars and were full of dust, just show them to people, and show how people used to live before. So it will be a diverse space with more facilities all our visitors can enjoy. The shop was finished in record time of just two months, including preliminary work in obtaining the necessary documentation to the last screw at the store. The shop is open all year, so we are always available to our customers.

The investment costs for the new organic shop are:

[T1] Investment construction costs for bio shop = 22,458 €

[T2] Costs of material during the construction of bio shop = 5544.88 €

[T3] Costs of interior design including installation = 10,313 €

starting from project documentation and geodetic study to the creation of all the necessary work finalization of the new shop.

Total investment costs (TIC)

$$\begin{aligned} \text{TIC} &= \text{T1} + \text{T2} + \text{T3} \\ \sum(\text{TIC}) &= 38315.88\ € \end{aligned} \tag{5}$$

4.1.1. Production plan for sales in the new organic shop

Production plan is as realistic as possible with the planned annual quantities in order to treat them as further planned business activities and investment for our new shop, the place where we are going to sell our organic products. This kind of business philosophy is more known as direct sales. Our organic farm lives by a system of family-run business, and according to this statement, we have human and natural limits of the production on the island. Nature is generous but it has limitations. It is an unwritten rule that depending on how much we give the nature, this much she gives back to us. Organic production is divided into six groups: honey, immunity products, olive oil, natural cosmetics, noble drinks, and island delicacies. Our farm, with its space and technology, integrates the process of gathering, enriching, and packing honey and other agricultural produces. Our production standards are realized through the hazard analysis and critical control point (HACCP) program, and the production object is the export object of the final product in the European Union countries.

4.1.2. Honey

Beekeeping in our farm is based on mobile ecological beekeeping on the south part of the island of Rab, the Barbat village and the Gorski kotar region. From the beginning of our activities, we have an average of one hundred beehives. As part of our business philosophy, we produce rare highly aromatic honeys that with their direct bactericidal and medicinal effect take up the lead position in everyday alimentation. After the bee pastures of sage (*Salvia officinalis* L.) and other medicinal honey plants on the island of Rab, we transport our bees (late June) into the mountains, in the region of Gorski kotar where we have bee pasture in mountain meadows and spruce and fir, the forest honey. Taking the two different locations of our apiaries into account, we estimated the production quantities in 2015 to be sold in our organic shop on the island of Rab. The plan for sage honey is 280 kg. The plan for multifloral honey "Bilje Kvarnera" is 480 kg. The plan for the forest honey "Šuma Kvarnera" is about 600 kg. The quantities will be divided and distributed in jars of various sizes, ranging from 212 to 580 ml. Production costs with quantities for honeys are:

Honey Salvia (150 g) = 666 €

Honey Salvia (260 g) = 576 €

Honey Bilje Kvarnera (400 g) = 2800 €

Honey Bilje Kvarnera (780 g) = 1300 €

Honey Šuma Kvarnera (400 g) = 2800 €

Honey Šuma Kvarnera (780 g) = 1300 €

Total costs (Σ) = 9442 €

4.1.3. Immune system products

This group represents four products. It is a synthesis of multiple products from the beehives. When they are blended together, they achieve much better effect on the human body. There is a strong emphasis on sage as a natural antiseptic as well as other products such as propolis, which has a great antibacterial effect. These are the planned production quantities for 2015 and sale through our organic shop: "Salvia Immunity" 260 g, 700 pieces; "Salvia Protect" 30 ml, 800 pieces; "Abies Immunity" 260 g, 400 pieces; and "Propolis" 30 ml, 200 pieces. Production costs with quantities for this group of products are:

Abies Immunity (260 g) = 2000 €

Salvia Immunity (260 g) = 4900 €

Salvia Protect (30 ml) = 800 €

Propolis (30 ml) = 800 €

Total costs (Σ) = 8500 €

4.1.4. Olive oil

Olive oil has always been known for its medicinal characteristics. Regular and long-term usage of olive oil in our diets reduces the risk of many diseases [7]. For better understanding, here is the classification with description of olive oil.

The top quality olive oil is labeled like extra-virgin comes from virgin oil production only. The spiciness and bitterness sometimes seem too aggressive, but it is only a proof that the oil is rich in all those ingredients that have a beneficial effect on health and which are a waste to lose through cooking or frying [8].

Virgin olive oil comes from virgin oil production only but is of slightly lower quality with free acidity of up to 1.5% and is judged to have a good taste but may include some sensory defects [9]. Olive pomace oil is refined pomace olive oil often blended with some virgin oil. It is fit for consumption, but may not be described simply as olive oil. It has a neutral flavor and also a high smoke point [9].

Natura Rab has only premium quality extra-virgin olive oil, usually with oleic acids below 0.5%. In our plantations you can find different olive varieties, like "Oblica," "Levantinka," "Leccino," and "Pendolino." Olives are harvested on a daily basis by a handpicking system and immediately transported into the mill for processing into the finest olive oil. Almost all annual quantity of olive oil is sold through direct sale on our farm. In 2015, we plan to have 250 l of extra-virgin olive oil. This quantity we want to divide in 150 0.75 l bottles and 50 2 l bottles of. Production costs with quantities for olive oil are:

Olive oil (2 l) = 850 €

Olive oil (0.75 l) = 1400 €

Total costs (Σ) = 2250 €

4.1.5. Natural cosmetics

The richness of the Rab archipelago in over 800 herb species is a source of the island's aromatherapy. Some of them we use in production of our natural cosmetic line. We produce creams, oils, and soaps. The result of the continuous development in this sector is reflected by the entry in the register of the open cosmetic manufacturers. It is our duty but also a further confidence for end consumers of our products. For production, we use natural resources that grow on our organic farm, such as organic beeswax and extra-virgin olive oil. There are three main plants we use for the production of our cosmetic products. These are St. John's wort, immortelle, and lavender. These are the planned production quantities for 2015 for further sale in our organic shop: "St. John's wort" cream 50 ml, 300 pieces; "St. John's wort" oil 100 ml, 300 pieces; "Imortelle" cream 50 ml, 500 pieces; "Imortelle" oil 100 ml, 400 pieces; "Lavender" cream 50 ml, 200 pieces; and "Lavander" oil 100 ml, 200 pieces. In 2015, we produced 600 pieces of natural soaps, random kinds. Production costs with quantities for natural cosmetics are:

Imm. cream (50 ml) = 3000 €

St. John's wort cream (50 ml) = 1500 €

Lav. cream (50 ml) = 900 €

Imm. oil (100 ml) = 2200 €

St. John's wort oil (100 ml) = 1500 €

Lav. oil (100 ml) = 800 €

Gentle soaps (100 g) = 1200 €

Total costs (Σ) = 11,100 €

4.1.6. Brandies and liqueurs

According to the rich family tradition, we produce three types of our local brandies made with grape brandy as basis. The first one is the popular medica, homemade herb-flavored brandy. The second one is the fig liqueur, pure nature and phenomenal taste for someone who likes sweeter drinks. It is macerated organic dried fig in grape brandy. The third one, honestly, requires the least work in the production, but it does not mean that it is less valuable. It's called "Ruta" (*Ruta graveolens* L.); it got its name from the medicinal herb that is the main ingredient of this brandy. For all three brandies, we have one rule. It is a great experience to drink it out of a clay bićerin (small brandy glass) as an aperitif but also as a digestive. The special taste remains if it is drunk well chilled. These are the planned production quantities for 2015: "Travarica i eko med" 0.5 l, 400 pieces; "Smokovača" 0.5 l, 500 pieces; and "Ruta" 0.5 l, 250 pieces. Production costs with quantities are:

Medica (500 ml) = 2400 €

Smokovača (500 ml) = 3500 €

Ruta (500 ml) = 1000 €

Total costs (Σ) = 6900 €

4.1.7. Island delicacies

Island delicacies represent products like fig jam, lemon jam, organic honey vinegar, and organic olives in brine. These are the products that are created by a long-based family tradition of preparing natural food. Some of them are made only on demand (special orders or business gifts), while most of them are constantly available in our shop. These are the planned production quantities for 2015: organic honey vinegar 0.5 l, 400 pieces; olives 370 g, 200 pieces; fig jam 630 g, 300 pieces; lemon jam 630 g, 200 pieces; and honey biscuits 200 g, 300 pieces. Production costs with quantities are:

Honey vinegar (500 ml) = 1000 €

Olives in brine (380 g) = 500 €

Fig jam (630 g) = 900 €

Lemon jam (630 g) = 400 €

Honey biscuits (220 g) = 600 €

Total costs (Σ) = 3400 €

4.1.8. Variable costs and fixed costs on an annual basis in organic shop

Variable costs:

—Tasting the products on an annual basis = 800 €

—Cardboard packaging (bags and other supplies) = 1200 €

—Maintenance and cleaning = 500 €

—Energy (electricity, water) = 420 €

Total = 2720 €

Fixed costs:

—Promotional material = 300 €

—Costs of salesperson in the shop = 4500 €

—Insurance of shop (fire, earthquake, theft) = 100 €

Total = 4900€

4.1.9. Planned income cash in the new organic shop on an annual basis

Total investment costs = 38315.88 €

Total production and sale costs = 49,212 €

TR = 62,500 €

FR = TR – TC

FR = 13,288 €

4.2. Project 2: investment in olive plantation

If we look from the perspective of agriculture on the Croatian islands, specifically on the island of Rab, there is a problem with the available land for cultivation. There are many reasons for that: smaller areas, difficult access to fields (sometimes just on foot), unsorted fields with undivided ownership, etc. In our case, it requires constant investments of new plantations with typical plants for our area. Maybe it is weird but it is certainly true; the ratio between islands and mainland is as follows: a thousand square meters land on the island is like one hectare of land on the mainland. Generally, the Mediterranean plant species such as olives, figs, lemons, and vine are grown. Agricultural land in which we would like to invest is located on the island a few kilometers away from our organic farm, and the surface is 2500 m^2. The land is a family legacy, and on it are already five adult olive trees. There is a place for more trees, and we decided to plant new 18 olive trees of our typical varieties. For better understanding, planting of olive trees is the easy part of the project. Before that, the land has to be prepared and protected against external factors, so we decided to make a 40-meter stone wall and 150-meter long fence. We also have to ensure an agricultural water connection for the irrigation system. Later, when the whole project is completed, crops should be maintained. According to our estimates, this location should give in their full fertility about 1200 kg of olives per year in ideal conditions. With the implementation of this project (calculated total costs of investment is 6620 €), we increase the annual production of olive oil, but also the work volume increases. Under "construction costs of the new plantation, surface 2500 m^2," the types of costs stated below have been taken into account:

—Excavation and cleaning channel

—Soil preparation and transport of biowaste

—Soil melioration with alignment of land

—Digging holes for seedlings with bagger 1 × 0.8 m

—Making of the main entrance gate

—Making of supporting stone wall

—Agricultural water connection

—Making of 150 m protective fence around the property (two persons)

The cost of the new investment plan we want to restore with picking up the new olives and transforming them into our product, the extra-virgin olive oil (production and total revenue based on 18 olive trees). From experience and knowledge in agriculture, in the first 4 years, we cannot count on the return of the investment because there is no cash flow from selling the products. Trees are too young, and the first crops will be available in the fifth year of growth.

Calculation scheme:

Total costs (variable + fixed costs) = 558 €

N = 10 years

TC = 58.8 €/year

TC = 976.65 + 58.8 €

TC(y) = 1035.45 €

CP = total costs/yield

CP = 1035.45 €/1260 kg

CP = 0.82 €/kg olives

100 kg olives = 16 l olive oil

6.25 kg = 1 l olive oil

CP = 0.82 € × 6.25 kg

CP = 5.14 €/L

Price of 5.14 € is the breakeven for 1 l of olive oil based on the price of 0.82 €/1 kg olives. This input data is used in the next calculation for olive oil production (see **Table 2**).

Years	Yield (kg)	Olive oil production 6.25 kg/1 l	Cash flow through years €20/l
1–4	0	0	0
5	180	28.8	576
6	360	57.6	1152
7	720	115.2	2304
8th and further years (a 70-kg tree)	1260	20.6	4032

Table 2. Production and total revenue based on 18 olive trees (reproduction 16%).

In intensive plantations, the life of olive trees spans to about 50 years and can be divided into several periods. The nonproductive period is until the end of the fourth year. From sixth to seventh year is the period of initial cropping. From 8 to 30 years (the most important period) is a period of full fertility and economic standpoint. Variable costs for olive oil production from the eighth year onward including costs such as extra-virgin olive oil, chemical and sensory analysis, charge costs of oil bottle "Dorica" 750 ml, labels, PVC caps, plastic screw cap, and dispenser 31 × 24 mm is 1355.98 €.

Planned annual cash flow from sales of olive oil 0.75 l is calculated as:

CP = TC/Y

CP = 1355.98 €/201.6 l

CP = 6.73 €/L

FR = TR – TC

FR = 4020 € – 1355.98 €

FR = 2664.02 €

4.3. Project 3: beekeeping investment

In 2015 we decided to invest in the field of beekeeping. Optimization of transport resources is solved with purchase of a new work vehicle, which is also the largest investment this year in beekeeping (total investment costs = 18428.50 €). Of course, a multipurpose vehicle has more functions, so it will be used for other agricultural works as well. Also, we made a decision to widen the existing apiary. Therefore, we bought new wooden beehives which will be applied when necessary. There is also other necessary professional equipment and tools which have to be changed, some of it at short-time intervals, some of them not so often. The maintenance in beekeeping has to be mentioned, because it represents a large item in annual costs, like the maintenance of trailer, truck, and others. It should be noted that beekeeping demands the greatest amount of work and time on our farm, and consequently, the whole family is involved in beekeeping activities. In recent years, we have invested significantly in the field of beekeeping, but nature is very unpredictable. It is normal to always hope for the best, but the human factor is not the most relevant here. So, we cannot be sure about the quantities of honey products on an annual basis. This year's investments are analyzed through the planned revenue from the sales of bee products through our three sales channel, such as export through distributers, Web shop, and fairs which presents total income cash = 66,124 €, considering all cost parameters (variable outdoor production costs = 8562 € and variable costs of indoor production and sale = 16024.80 €). Particularities of our micro-region are always represented through the sales of our organic products, and we always emphasize the relationship between possibilities and realities in nature.

Estimated annual cash flow through three sales channels is calculated as:

Total revenue = 66,124 €

Total costs = 24586.80 €

FR = TR − TC

FR = 66,124 € – 24586.80 €

FR = 41537.20 €

4.4. Financial (CBA) analysis of individual investment projects

4.4.1. New organic shop

—Investment value of the new organic shop is 38315.88 €.

—Total production and sale costs are 49,212 €.

—Planned annual base income cash in bio shop is 62,500 €.

—Financial result for this project is 13,288 €.

—Net present value is 8260.55 € at the discount rate of 5.5%.

—Repayment period of investment is in year 4, where Investment flow is 0.21€ (**Table 3**).

4.4.2. New olive plantation

—Investment value of new olive plantation is 6620 €.

—Total production costs are 1355.98 €.

—Annual income cash is 4020 €.

—Financial result for this project is 2664.02 €.

—Net present value is 567.35 € at the discount rate of 5.5%.

—Repayment period of investment is in year 3, where Investment flow is 0.36 € (**Table 4**).

Year	Discount rate 14.51%	NPV-investment flow
1	11604.23	–26711.65
2	10133.81	–16577.84
3	8849.72	–7728.13
4	7728.34	**0.21€**
5	6749.05	6749.26
6	5893.85	12643.11
7	5147.02	17790.13
8	4494.82	22284.95
9	3925.26	26210.21
10	3427.88	29638.09

Table 3. NPV assessment for project 1.

Year	Discount rate 10.04%	NPV-investment flow
1	2420.96	–4199.04
2	2200.07	–1998.97
3	1999.34	**0.36€**
4	1816.92	1817.28
5	1651.14	3468.42
6	1500.49	4968.91
7	1363.59	6332.50
8	1239.18	7571.68
9	1126.11	8697.79
10	1023.37	9721.16

Table 4. NPV assessment for project 2.

4.4.3. Beekeeping investment

—Investment value for project 3 is 18428.50 €.

—Total production and sale costs through three channels are 24586.80 €.

—Planned annual base income cash for project 3 is 66,124 €.

—Financial result for project 3 is 41537.20 €.

—Net present value is 20943.25 € at the discount rate of 5.5%.

—Repayment period of investment is in year 1, where Investment flow is 0.53 € (**Table 5**).

Year	Discount rate 125.39%	NPV-investment flow
1	18429.03	**0.53€**
2	8176.51	8177.04
3	3627.72	11804.76
4	1609.53	13414.29
5	714.11	14128.40
6	316.83	14445.23
7	140.57	14585.80
8	62.37	14648.17
9	27.67	14675.84
10	12.28	14688.11

Table 5. NPV assessment for project 3.

5. Conclusion

The aim of this research was to study the economic validity of the three projects, considering economic parameters for the return of the investment (the net present value and the internal rate of return) and my input information. A model was developed in Microsoft Excel for the net present value assessment, which serves as a support for decision-making, should we go into the investment or not, or better said does the investment make sense, and when we expect the return in the terms of time.

Using the NPV and IRR methods showed, the return of the investment into new organic shop will be in 4 years. The return of the investment into new olive plantation will be in 3 years. The last investment into new beekeeping facilities will be by 5.5% discount rate in the first year (NPV = 20943.25 €).

With the implementation of these projects, we wanted to optimize our business resources, and in terms of productivity, we are much better than before. The projects have enabled easier performance in our obligations, but consequently, they increased the workload in both parts, first in the sales (project 1) and in terms of production (projects 2 and 3).

Author details

Karmen Pažek, Matija Kaštelan, Martina Bavec, Črtomir Rozman and Jernej Prišenk*

*Address all correspondence to: jernej.prisenk@um.si

Faculty of Agriculture and Life Sciences, University of Maribor, Hoče, Slovenia

References

[1] Turk, J. Agricultural economics: theory and aplication. Maribor: Faculty of Agriculture and Life Sciences; 1998. 176 p.

[2] Gittinger, J.P. Economic analysis of Agricultural Projects [Internet]. 1985. Available from: http://www-wds.worldbank.org/external/default/WDSContentServer/WDSP/IB/2000/04/19/000178830_98101903531847/Rendered/PDF/multi0page.pdf [Accessed: 2015-09-28]

[3] Turk, J. Theoretical and empirical analysis in agricultural economics. Maribor: Faculty of Agriculture and Life Sciences; 2001. 225 p.

[4] Berk, J., De Marzo, P., Stangeland, D. Corporate Finance (3rd Canadian ed.). Toronto Pearson, Canada; 2015. 64 p.

[5] Law, A.M. Using net present value as a decision-making tool. Air Medical Journal . 2004;23(6):28–33. DOI: 10.1016/j.amj.2004.08.025

[6] Wikipedia. Net present value [Internet]. 2016. Available from: https://en.wikipedia.org/wiki/Net_present_value [Accessed: 2016-07-26]

[7] Covas, M. Olive oil and the cardiovascular system. Pharmacological Research. 2007;55:175–186. DOI: 10.1016/j.phrs.2007.01.010

[8] Eko centar Natura Rab. Natura Rab [Internet]. 2016. Available from: http://www.natura-rab.hr/en/products.html [Accessed: 2016-07-28]

[9] International Olive Council. Designations and definitions of olive oils [Internet]. 2015. Available from: http://www.internationaloliveoil.org/estaticos/view/83-designations-and-definitions-of-olive-oils [Accessed: 2015-09-10]

Chapter 7

Influence of Phosphorus Precipitation on Wastewater Treatment Processes

Ján Derco, Rastislav Kuffa, Barbora Urminská, Jozef Dudáš and Jana Kušnierová

Additional information is available at the end of the chapter

http://dx.doi.org/10.5772/65492

Abstract

Phosphorus stimulates aquatic plant growth and contributes to eutrophication process in rivers, lakes and the ocean. A large part of phosphorus is discharged into the receiving waters by wastewater. One of the solutions of this problem is represented by chemical precipitation. Simultaneous precipitation of phosphorus from wastewater with metal salts is commonly applied. Metal salts are dosed directly into aeration tank, and produced precipitates are wasted as a part of the secondary sludge. Thus, not only aerobic and anoxic processes of wastewater treatment plant are carried out in the presence of precipitant metals and precipitates but also the precipitates are, in many cases, accumulated in anaerobic sludge digesters. Operational research of phosphorus precipitation in lab-scale encompasses the impact of Fe^{2+}, Fe^{3+} and Al^{3+} salts on biological nitrification and denitrification processes, sedimentation and thickening characteristics of sludge as well as anaerobic sludge stabilisation processes. The measurements of specific oxygen uptake rate, nitrification and denitrification tests and monitoring of effluent values of quality standards were applied to evaluate the processes performance. Other objective of our research is to contribute to methodology for examination of thickening and dewatering characteristics of sludge with tested precipitation agents. Mathematically processed experimental results are used to compare sedimentation, precipitation and dewatering characteristics of activated sludge cultivated in the presence of selected precipitation agents. Better description of the experimental results was obtained with three parameters model of particles mass flow density curve. Comparison of minimum sedimentation tank size necessary for gravitational separation of individual sludge was used to examine sedimentation characteristics of activated sludge. Thickening characteristics of sludge were evaluated based upon thickening area needed to maintain the required sludge concentration in activated sludge model, which corresponds to maximum surface load in undissolved substances. Chemical precipitation of phosphorus produces metal precipitates. These are transported with the waste sludge to the digestion tanks. Impact of the precipitates on the anaerobic sludge stabilisation process as well as on the sludge water quality was tested in the work. The main aim of the research and proposed chapter submission is a pursuit of decision making regarding selection of type of precipitating agent and strategy of chemical phosphorus precipitation.

Keywords: aerobic, aluminium salts, anaerobic digestion, anoxic, activated sludge, denitrification, dewatering, excess sludge, ferric and ferrous precipitants, inhibition, heterotrophic microorganisms, nitrification, operating research, plug flow, respirometric measurements, simultaneous phosphorus precipitation, sludge water, solids flux flow, specific oxygen uptake rate, stimulation, thickening

1. Introduction

The requirement for removal of phosphorus from wastewater (WW) is rising due to increasing problem of eutrophication all over the world. Biologically enhanced phosphorus removal represents the most convenient way. The main disadvantage of this approach is the requirement for an anaerobic bioreactor, that is, an increase of capital costs. As a result, chemical precipitation with ferric, ferrous, aluminium or calcium salts is usually applied. There are various technological strategies (**Figure 1**), which depend on the place of precipitants dosing with regard to the position of biological stage of the wastewater treatment plant (WWTP). Other reason for chemical phosphorus removal can be temporary solution during upgrading of a WWTP for enhanced biological phosphorus removal. There are also circumstances in WW treatment practice (e.g. lower organics content in WW for sufficient biological phosphorus removal, lower winter temperatures or more stringent effluent phosphorus standards) for application of chemical phosphorus precipitation as a complementary solution. Combined chemical and biological phosphorus removal is more effective and chemical saving.

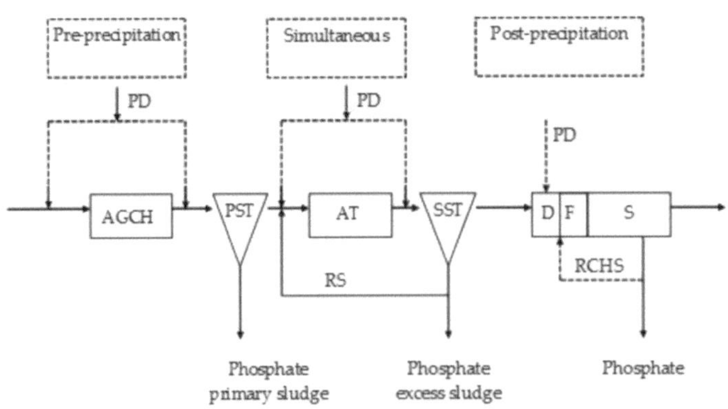

Figure 1. Phosphate removal processes (PD—precipitant dosing, AGCH—aerated grid chamber, PST—primary sedimentation tank, AT—aeration tank, SST—secondary sedimentation tank, D—dosing, F—flocculation, S—sedimentation, RCHS—returned chemical sludge).

http://dx.doi.org/10.5772/65492

The addition of chemicals to raw wastewater for precipitation of phosphorus in primary sedimentation facilities is termed *"pre-precipitation"* [1]. Precipitates of phosphorus are withdrawn from the system as a part of primary sludge with low effect on activated sludge microorganisms. Increase in the amount of primary sludge is typical for this treatment technology. Precipitants are helpful to eliminate fluctuations of organic mass loading to biological stage of WWTP. However, simultaneous coagulation of organic matter occurs, leading to a decrease in BOD_5 and possible negative impact on efficiency of denitrification process.

Simple implementation of precipitating reagent dosing but no recovery of phosphorus from primary sludge is characteristic for this strategy.

The addition of chemicals to form precipitates that are removed along with wasted biological sludge is defined as *"coprecipitation"* [1]. Chemicals can be added to the effluent from primary sedimentation tank—the mixed liquor, that is, directly to the biological stage, or to the effluent from biological treatment process before secondary sedimentation.

Enhanced sedimentation of activated sludge is the advantage of phosphorus *coprecipitation* or *simulated precipitation*. Very simple dosing of precipitating chemicals and sufficient mixing and flocculation in biological treatment facilities are typical for this strategy. Concentration of mixed liquor in biological treatment tank increases, and solid retention time (SRT) decreases as a consequence of phosphorus precipitation. Phosphorus precipitates are removed from the treatment system as a part of excess sludge from which the phosphorus is not recoverable. Decrease in acid neutralisation capacity and pH in biological treatment tank and direct impact of precipitant agents on activated sludge microorganisms are also typical features of this phosphorus removal strategy.

"Post-precipitation" involves the addition of chemicals to the effluent from secondary sedimentation facilities and the subsequent removal of chemical precipitates. In this process, the chemical precipitates are usually removed in a separate sedimentation facilities or in effluent filters [1]. For this strategy, the elimination of effect of reagents on activated sludge microorganisms is characteristic. Main advantage is the achievement of minimal concentration values of total phosphorus in effluent (1 mg L^{-1}). Phosphorus is recoverable from chemical sludge. Main disadvantage of this technology is high investments related to separate flocculation and sedimentation tanks.

Simultaneous precipitation of phosphorus from wastewater with iron and aluminium salts is commonly used. Precipitation agents are dosed directly into the aeration tank, and produced precipitates are wasted as a part of the secondary sludge. Thus, not only aerobic and anoxic processes at WWTP are carried out in the presence of heavy metals, but precipitates are in many cases accumulated also in anaerobic sludge digester.

The most important operating parameter of simultaneous phosphorus removal is precipitating agent dosage. A lot of information about the dosing of precipitants can be found in the literature. Dosage of aluminium and iron salts usually falls into the range of 1–3 metal ion/phosphorus on a molar ratio basis [Eq. (1)] if the residual phosphorus in the secondary effluent is >0.5 mg L^{-1}. To achieve phosphorus levels below 0.5 mg L^{-1}, significantly higher metal salt doses and filtration will be required [1]:

$$\beta = \frac{n_{Me}}{n_P} \tag{1}$$

where n_{Me} is the molar concentration of precipitating agent required for precipitation [mol L^{-1}], n_P the molar concentration of phosphorus to be precipitated [mol L^{-1}].

Recommended values of β ratio to achieve effluent concentration of phosphorus below 1.9 mg L^{-1} are 2–4 in simultaneous precipitation of phosphorus [2]. Under optimal function of separation stage, the concentrations of Al^{3+}, Fe^{2+} and Fe^{3+} in the effluent discharge for B = 1.5 are lower than 1 mg L^{-1} [3], that is, often lower than in the raw water. The separation by filtration enables to achieve metal concentrations below 0.5 mg L^{-1}. For urban wastewater with an average concentration of chloride 100 mg L^{-1}, sulphates 200 mg L^{-1} and the total phosphorus 10 mg L-^{1}, an increase in chloride concentration by about 50%, respectively 25% for the sulphate, will be the consequence of chemical precipitation process. It is also apparent that precipitation of phosphate contributes to the reduction in the total concentration of dissolved inorganic salts [4].

The required dose of coagulant/precipitant depends on the phosphate concentration, the pH and composition of the water. Due to the varying flow and phosphorus concentrations in raw water, it is advantageous to optimise the dosing of coagulants experimentally.

Significantly less data for the influence of precipitants on activated sludge activity and process efficiency appear in the literature. Mowat [5] published the data related to metal toxicity on microorganisms during biochemical oxygen demand test. At 20 mg L^{-1}, aluminium corresponded approximately to trivalent chromium and cyanide, and ferric iron showed similar toxicity. According to Bever and Teichman [3], ferric iron has a stimulation effect on nitrification process but ferrous iron causes an inhibition of this process. but inhibition of this process by ferrous iron was published by Bever and Teichman [3]. Because iron is usually introduced as not very toxic metal, there is a scarcity of literature about its toxicological and inhibitory effects on freshwater organisms. Aluminium and iron are commonly bioaccumulating in their salt form. Aluminium is a non-essential element, and it is mainly discussed in the context of its suspected detrimental role on the uptake of essential elements [6].

Extending spectrum of applied wastewater treatment processes and technologies follows the ever stricter requirements for discharged wastewater quality. Chemical precipitation of phosphorus is applicable in municipal wastewater treatment, and although being less environmentally favourable compared to enhanced biological phosphorus removal, the amount of investment funds is often decisive at this stage. Furthermore, enhanced biological phosphorus removal process frequently requires the application of chemical precipitation/post-precipitation as well, taking the increasingly stringent requirements for phosphorus removal from wastewater into account.

http://dx.doi.org/10.5772/65492

Simultaneous chemical phosphorus precipitation in biological stage of WWTP represents one of the most commonly used technologies when Fe^{2+}, Fe^{3+} and Al^{3+} salts are dosed into biological treatment facilities to precipitate the excess phosphorus in wastewater. These salts and their precipitates then become a part of the excess sludge and are thus transported also into anaerobic stabilisation tank.

Similarly to most treatment processes, chemical phosphorus precipitation is accompanied with generation of side products, interactions and impact on other simultaneously running processes. The aim of our research was to compare the impact of Fe^{2+}, Fe^{3+} and Al^{3+} salts on sedimentation and thickening characteristics of sludge, biological transformation processes and ammonia nitrogen removal, that is, nitrification and denitrification processes, as well as anaerobic sludge stabilisation processes. The objective of our research is to contribute to methodology for examination of thickening and dewatering characteristics of sludge with investigated precipitation agents. Mathematically processed experimental results of lab-scale-operated activated sludge models are used to compare sedimentation, precipitation and dewatering characteristics of activated sludge cultivated in the presence of selected precipitation agents, that is, Fe^{2+}, Fe^{3+} and Al^{3+} salts. Other purpose of this study is to find methods for quantification of precipitants influence on activated sludge activity.

2. Experimental and data processing methods

2.1. Influence of aerobic and anoxic biological processes

The process of simultaneous precipitation of phosphorus in aerobic biological WWTP was simulated in lab-scale. The activated sludge was cultivated in semicontinuous reactors in order to simulate plug-flow hydraulic regime at real aerated tanks of municipal WWTPs. The volume of one unit was equal to 1.5 L. The cultivation was carried out by applying different synthetic wastewater composition, that is, an individual substrate (methanol) or mixed substrates (methanol, glucose, natrium acetate and peptone or the mixture of natrium acetate and glucose) and ammonium nitrate in accordance with the ratio BOD_5:N = 100:5.

Operational conditions were changed during the experiments in order to investigate also the influence of these variables on activated sludge activity. The volumetric load related to COD varied between 1.0 and 2.0 kg m^{-3} d^{-1}. The bioreactors were operated over the solid retention time (SRT) of 8 or 10 days. The hydraulic retention time (HRT) of 46 h was maintained in the bioreactor. The amount of phosphorus added into the systems corresponded to its effluent concentration approximately 9.0 mg L^{-1}. Ferrous, ferric and aluminium salts were dosed as phosphorus precipitants and were applied after 50 days of cultivation (activated sludge acclimation). The amounts of precipitants required for dosing were obtained in accordance with metal/phosphorus molar ratios ranging from 1.5 to 3.0. The alkalinity was adjusted with sodium hydrocarbonate solution.

The analysis of chemical oxygen demand (COD), total suspended solids (TSS), volatile suspended solids (VSS) and soluble phosphorus content was performed in accordance with procedures described in the standard methods [7]. The content of total organic carbon (TOC) was obtained by TOC 2000 P analyser produced by IPU.

Respirometric measurements of oxygen uptake rate (OUR) which were performed by an oxygen Syland probe evaluated the influence of the above given precipitants on the activated sludge activity. These measurements were performed according to [8–10] and were carried out before the new aeration cycle started, that is, with endogenous biomass. With the resulting values, the maximum total oxygen uptake rates (the sum of exogenous and endogenous rates) were obtained. The effect of precipitants on activated sludge was evaluated from the total specific oxygen uptake rate (SOUR) values of metal-laden system related to the total SOUR values of the control system (without metals exposition). Thus, stimulation effect of metal to the activated sludge is calculated as follows:

$$SE = \left(\frac{r_{x,t}}{r_{x,t,con}} - 1 \right) . 100 \ [\%] \tag{2}$$

where SE is the stimulation effect [%]; $r_{x,t,con}$ the total specific oxygen uptake rate of control sludge [mg g^{-1} h^{-1}]; $r_{x,t}$ is the total specific oxygen uptake rate of heavy metal-laden sludge [mg g^{-1} h^{-1}].

The measurements of specific substrate removal rate R_X at various substrate concentrations were performed by respirometric method [8, 9], and the values of maximum respiration rate $R_{x,max}$ and half saturation constant K_s of the Monod equation:

$$R_X = R_{x,max} \frac{S}{K_S + S} \tag{3}$$

where K_S is the half saturation constant [mg L^{-1}], R_X is the specific rate of substrate removal [mg g^{-1} h^{-1}], $R_{X,max}$ is the maximum specific rate of substrate removal [mg g^{-1} h^{-1}], S is the substrate concentration [mg L^{-1}],

were also evaluated in order to compare the influence of precipitants on activated sludge activity. Grid search method [11–13] was applied to evaluate respirometric measurements carried out at different substrate concentrations.

2.2. Impact on sedimentation, thickening and dewatering properties

Monitoring the impact of phosphorus precipitation agents to sedimentation, thickening and filtration characteristics of activated sludge formed another part in our research of the impact of precipitation agents to biological wastewater treatment processes and sludge treatment.

Real activated sludge process performed in systems with concentration gradient was simulated in the lab in semi-continuous models operated with default sludge age of 10 days. Retention time of synthetic wastewater was 2 days, and organic load rate expressed as COD was 1.5 kg $m^{-3}\,d^{-1}$. Salts of Fe^{2+}, Fe^{3+} and Al^{3+} were dosed to the systems together with the substrate. Fe was added as $FeSO_4{\cdot}7H_2O$. Trivalent Fe was added only after $FeSO_4{\cdot}7H_2O$ oxidation. Al^{3+} salts were added in form of $AlCl_3{\cdot}6H_2O$. Metals doses to precipitate excess phosphates were calculated for $\beta = 1.5$. A control model was operated in parallel, that is, without precipitation agents. Synthetic wastewater contained glucose and sodium acetate (COD = 3000 mg L^{-1}).

Solid flux method [1, 14–16] for the analysis of settling/thickening data and filtration equation [17–19] for processing of dewatering data were applied. Respiration measurements of SOUR were performed in order to evaluate the influence of precipitating agents on activated sludge microorganism activity.

In the case of non-standard sludge/suspension, it is possible to define the thickening area from experimental curve of particles mass flow density [18] that defines the relation of particles mass flow density q [kg $m^{-2}\,h^{-1}$] and sludge concentration [Eq. (4)]:

$$q = X \, . \, u_0 \, . \left(1 - \frac{X}{c_k}\right)^n \tag{4}$$

Particles mass flow density values may be gained as a product of sludge concentration values and respective measured precipitation rate or free sedimentation [15]. Analogically, the below formula applies for thickening rate u_z at specific solids concentration X:

$$u_z = u_0 \, . \left(\left(1 - \frac{X}{c_k}\right)^n\right) \tag{5}$$

where X means sludge concentration [kg m^{-3}]; u_0, c_k and n are empirical parameters; c_k parameter value characterises the given suspension; it represents the maximal solids concentration (compression region). The constant u_0 represents thickening rate at unit suspension porosity. Simplified two parameters formula below may be used to describe the dependence of thickening rate and sludge concentration [18]:

$$u_z = k \, . \, e^{-\beta . X} \tag{6}$$

The values of k [m h^{-1}] and $ß$ [$m^3\,kg^{-1}$] parameters describe the sludge thickening characteristics. Required thickening area may then be calculated from the equation:

$$A_Z = \frac{Q_0 \,.\, (1+R) \,.\, X_A}{q_{\min}} \tag{7}$$

where A_z means the required thickening area of thickening tank [m^2]; Q_0 is wastewater flow [$m^3\,h^{-1}$]; X_a means concentration of solids [$kg\,m^{-3}$]; and R means return sludge recirculation ratio. Minimum particles flow density q_{min} may be defined graphically pursuant to Yoshioka et al. [20] or calculated for example from the equation:

$$q_{\min} = \frac{u_0 \,.\, X_k \,.\, X_r}{X_r - X_k}\left(1 - \frac{X_k}{c_k}\right)^{-n} \tag{8}$$

If simplified two parameters thickening rate model is used [Eq. (6)], the below will apply for minimum flux of solids q_{min} (kg m^{-2} min^{-1}) moving downward:

$$q_{\min} = k \,.\, \beta \,.\, X_k^2 \,.\, e^{-\beta . X_k} \tag{9}$$

Tuček and Koníček [18] describe how the Eqs. (8) and (9) are derived, as well as the formulae to calculate critical concentration value X_k and returned sludge concentration X_r.

Thickening rate corresponding to the sludge concentration [Eq. (5)] needs to be higher than solid flux of tank, that is, the below applies for minimum thickening area $A_{Z,min}$:

$$A_{Z,\min} \geq \frac{Q}{u_z(X_a)} = \frac{Q}{u_0}\left(1 - \frac{X_a}{c_k}\right)^{n} \tag{10}$$

where Q means sludge flow rate [$m^3\,h^{-1}$]

Assuming that flow rate is laminar when the filtrate flows through the porous material, filtration process may be described by the equation [16, 17]:

$$\frac{dV}{S \,.\, d\tau} = \frac{\Delta p \,.\, S}{\alpha \,.\, \eta \,.\, C \,.\, (V + V_e)} \tag{11}$$

where Δp means overall pressure difference [Pa] prior and after the filter that needs to be developed in order the required filtrate flow rate is reached; S means the filter size [m^2]; dV means filtrate volume increment [m^3] in time dτ [s]; u means filtration rate [$m\,s^{-1}$]; α means specific cake resistance [m kg^{-1}]; η means dynamic filtrate viscosity [Pa s]; C means solids concentration in suspension [kg m^{-3}]; V means filtrate volume [m^3] in time τ [s]; and V_e means fictive filtrate volume [m^3] that would be developed by a cake with the same resistance as the filtrate material resistance.

Integrating Eq. (11) and other adjustments lead into Eq. (12) for filtration at constant pressure [18]:

$$\left(\frac{V_2 - V_1}{2}\right)\frac{2}{k} + \frac{2V_e}{k} = \frac{\Delta\tau}{\Delta V} \tag{12}$$

Characteristic filtration constants, that is, V_e and k quantities may be determined based upon measured values V_i, V_{i+1} and τ_i a τ_{i+1}. Then, the filtration cake-specific resistance may be calculated from Eq. (13):

$$k = \frac{2\,\Delta p\,.\,S^2}{\alpha\,.\,\eta\,.\,C} \tag{13}$$

Another approach to determine filtration cake-specific resistance value is based on the assumption of negligible filtration material resistance with regard to filtration cake resistance [15]:

$$\alpha = \frac{\Delta p\,.\,S^2}{\eta\,.\,C\,.V_1.\dot{V}_1} \tag{14}$$

Neglecting filtration support material resistance results in a relative error n defined as below:

$$n = \frac{\Delta\alpha}{\alpha}100 = \frac{\dot{V}_1}{\dot{V}_0 - \dot{V}_1}.100\ [\%] \tag{15}$$

where V_1 means the volume of filtrate used to create the cake; $\dot{V}_0$ means filtrate volumetric flow rate through the filtration material; and $\dot{V}_1$ means volumetric filtrate flow rate through the cake and filtration material.

The dependence of specific resistance of compressible cake from pressure difference may be expressed as [15]:

$$\alpha = \alpha_0 + a\,.\,\Delta p^x \tag{16}$$

For certain suspensions $\alpha_0 = 0$ and Eq. (16) shall thus be transformed to Eq. (17):

$$\alpha = a\,.\,\Delta p^x \tag{17}$$

where a is constant.

Mathematically, the constant a equals to specific cake resistance at unit pressure difference for Eq. (16). Parameter x represents the filtration cake compressibility rate; $x = 0$ for non-compressible cake. Mathematically, α_0 represents specific cake resistance at $\Delta p = 0$.

Sedimentation characteristics of sludge from individual systems were evaluated based upon the minimum thickening area $A_{Z,min}$ values [Eq. (10)] that correspond to those of maximum hydraulic surface load of sedimentation tank in order to ensure for sludge separation conditions.

Thickening characteristics of sludge were examined by comparing the thickening area values calculated for measured solids concentration and the same flow rates of wastewater and return sludge. Three parameters model of particles flow curve [Eq. (4)] was applied onto the results of thickening curves measurements, or the dependence of thickening rate from sludge concentration [Eq. (5)], and two parameters dependence of thickening curve from sludge concentration [Eq. (6)] was applied.

Experimental values of filtration characteristics (specific filtration cake resistance α and fictive filtrate volume V_e) determined in filtrate rate measurement at constant pressure difference were used to assess dewatering sludge characteristics. At the same time, α dependence from pressure difference was also measured.

Grid search optimisation method [10, 12] was used to determine parameters in Eqs (4), (6), (16) and (17). Residual sum square S^2_R between experimental and calculated q, u and α values was applied as objective functions [12]:

$$S^2{}_R = \frac{\sum (y_i^{\exp} - y_i^{cal})^2}{n - m} \tag{18}$$

where n represents the number of measurements and a means the number of model parameters.

2.3. Impact on anaerobic sludge stabilisation

Experimental models to simulate the processes occurring in the anaerobic digesters of stabilising raw sludge (**Figure 2**) consisted of three parts:

- reactor with volume 1.0 L which served as the digester with venting of generated biogas in the process,
- sorption of contaminants from produced biogas,
- device for measuring the resulting CH_4.

Reference model was also operated in parallel.

Anaerobically stabilised sludge sampled from the digester at WWTP Bratislava (applied as inoculum) was loaded with primary and excess sludge taken from the abovementioned WWTP. Anaerobic digesters were operated in batch mode at a residence time of 5 days. In both systems, during the tests carried out with the raw sludge the processes of the hydrolysis were investigated by measuring changes in the concentration of ammonia, and the methanogenesis process was investigated by measuring the production of biogas. Effect of iron on the processes of anaerobic digestion was evaluated by comparing the output values of the experimental (addition of Fe) and reference model. The pH was adjusted dosing systems $NaHCO_3$ and NaOH to the same value (about 7.8) to increase buffering capacity of both systems. The models were run at 37°C.

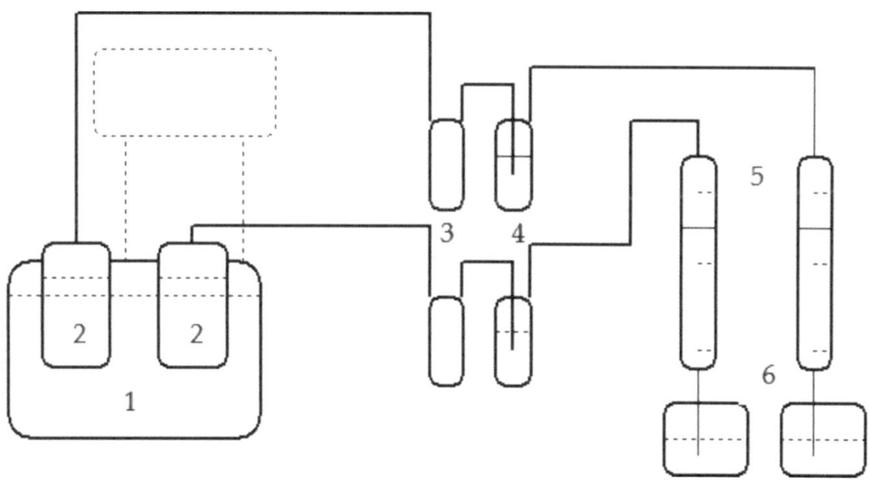

Figure 2. Schematic representation of lab-scale anaerobic stabilisation equipment (1-thermostat, 2—reactors, 3—safety vessels, 4—bubbler with KOH, 4—measurement of produced gas volume, 6—storage reservoirs for water).

3. Results and discussion

3.1. Impact on aerobic and anoxic biological treatment processes

The process of simultaneous precipitation of phosphorus in aerobic biological wastewater treatment system has been studied in semi-continuous lab-scale bioreactors. Operational conditions, type and dose of precipitants (ferric, ferrous and aluminium salts) were changed during experiments in order to investigate the influence of these variables on activated sludge activity. Short-term and long-term effects of precipitants on activated sludge activity have been studied. Respirometric measurements were performed in order to evaluate the influence of precipitant metals on activated sludge activity. The effect of precipitants was evaluated with regard to specific oxygen uptake rate (SOUR) of the control activated sludge system operated at the same conditions excluding the precipitants dosing.

Figure 3 shows the effect of ferrous ($FeSO_4 \cdot 7H_2O$) and aluminium ($Al_2(SO_4)_3 \cdot 18H_2O$) precipitants on activated sludge respiration activity. Model wastewater containing glucose, methanol, sodium acetate and peptone was used as a substrate. Bioreactors including control system were operating at SRT of 8 days and the volumetric load of 1.0 kg m^{-3} d^{-1}, related to COD. The precipitants during this set of experiments were dosed at metal to phosphorus molar ratios equal to 3.

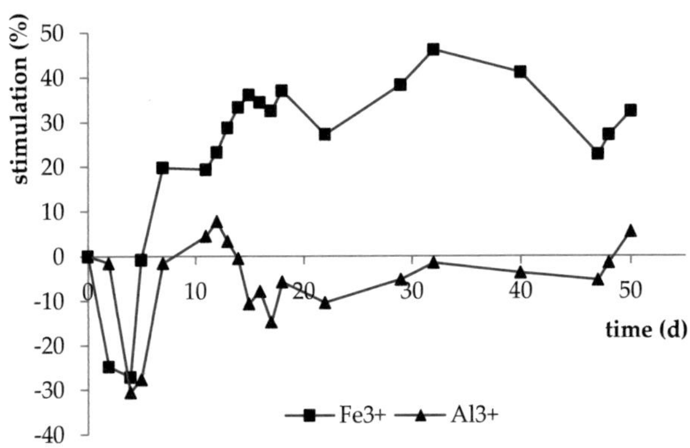

Figure 3. Stimulation effect of Fe^{2+} and Al^{3+} precipitating agents on activated sludge respiration activity.

The inhibition effect of about 30% for both precipitants few days after the beginning of the experiments follows from **Figure 3**. In the next period, after approximately 15 days from the beginning of the precipitants dosing, the stimulation effect of ferrous iron on activated sludge activity of about 30–40% [Eq. (2)] can be estimated from **Figure 3**. On the other hand, only small inhibition or insignificant stimulation influence of aluminium precipitant resulted from the measurements as can be seen in **Figure 3**.

The results on stimulation or inhibition effects of the precipitants on the activated sludge respiration activity can be influenced by the procedure applied in the evaluation.

From Eq. (2) follows that the values of relative SOUR can be considered to be a measure of these effects. The values of total suspended solids (TSS) are usually applied as a basis of specific rate expressions in wastewater engineering. It is obvious that the values of TSS concentration in the systems with precipitant dosing increase due to the precipitates formation and accumulation in activated sludge and will be higher in comparison with the control system.

http://dx.doi.org/10.5772/65492

Significant differences between the values of VSS versus the duration of experiments duration were observed in all three bioreactors maintained the same values of operational parameters (HRT, SRT, volumetric load) and used equal substrate composition [21]. These differences can be related for example to possible influence of heavy metals on biomass yield or to thermal decomposition of precipitates with hardly defined composition during performing the VSS content determination procedure. For example, the value of VSS for Fe precipitate based on our experiment is about 13% and for Al precipitate about 27% (both precipitates were prepared by coagulation, precipitation and filtration) [21]. It is also assumed that the differences between the VSS content in the studied systems can be partially ascribed to absorption of both dispersed form of biomass and slowly biodegradable products of microorganisms. This assumption follows from higher treatment efficiency of organic pollution removal in the systems with phosphorus precipitation with regard to dissolved as well as dispersed organic substances observed in our previous work [21]. Thus, adsorbed organics on activated sludge flocs practically decrease the observed values of activated sludge SOUR. This is due to the higher values of biomass concentration approximated by VSS content to which the values of SOUR are related.

The TOC content in biomass was also analysed in order to evaluate SOUR and compare the influence of the studied heavy metals on activated sludge activity. The values of TOC content in activated sludge from operated lab-scale bioreactors are given in **Table 1** [22].

Aerated system	TOC [%]
Fe^{2+}	17.10
Fe^{3+}	16.37
Al^{3+}	19.84
Control	21.33

Table 1. TOC content in activated sludge cultivated in observed lab-scale aerated system.

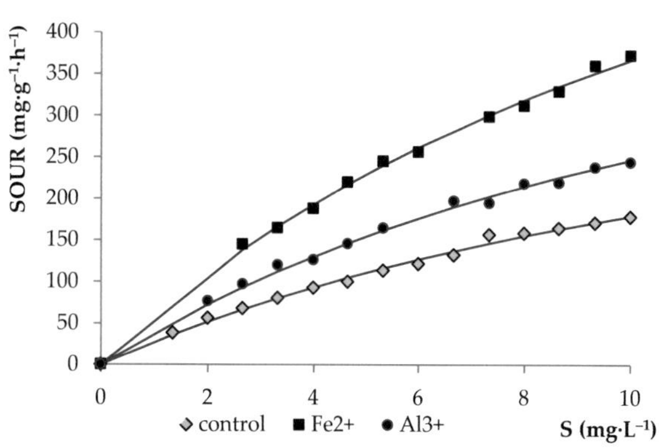

Figure 4. Dependence of SOUR (related to TOC mass unit) on substrate concentration.

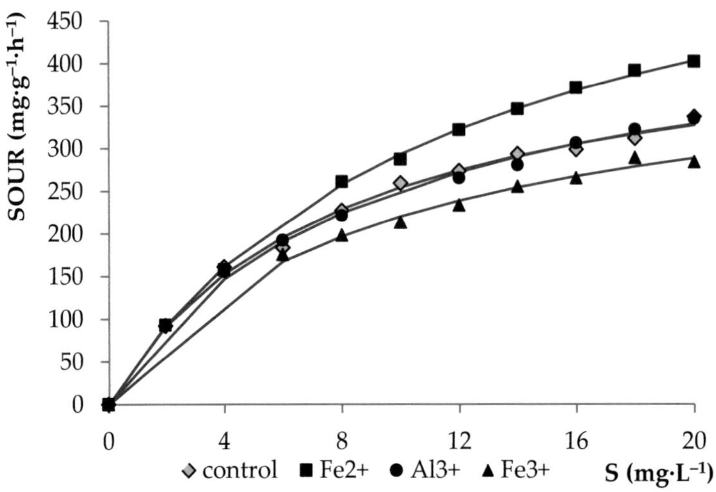

Figure 5. Dependence of SOUR (related to VSS mass unit) on substrate concentration.

The courses of the experimental and calculated values of SOUR (related to TOC concentration values) at different values of substrate concentration are presented in **Figure 4**. The highest values of SOUR can be concluded for the activated sludge cultivated in the presence of ferrous precipitant. On the other hand, this figure indicates only small differences between the SOUR courses obtained for activated sludge, which was cultivated in the control system and in the presence of ferric iron or aluminium precipitate. In **Figure 5**, similar concentration dependencies are plotted, but the values of SOUR for the same measurements are related to VSS content in individual bioreactors. Similarly to **Figure 4**, the highest values of SOUR have been achieved with activated sludge cultivated in the bioreactor with ferrous precipitant dosing. The values of Monod equation parameters obtained by the evaluation of SOUR values shown in **Figure 4** are given in **Table 2**.

	Control	Al^{3+}	Fe^{2+}	Fe^{3+}
$R_{X,max}$ [mg g^{-1}MLSS h^{-1}]	457.3	476.3	644.8	417.9
$R_{X,max}$ [mg g^{-1} TC h^{-1}]	1176.4	1326.6	2031.4	1252.1
K_S [mg L^{-1}]	10.0	10.0	15.0	10.0
$R_{X,max}$ [mg g^{-}1TOC h^{-1}]	2553.0	2533.3	4091.3	2659.1

Table 2. The values of biokinetic parameter values [Eq. (3)] related to different forms of activated sludge in aerated systems.

The courses of calculated values of SOUR with the same parameter values are also presented in **Figure 4**. A very good agreement between the experimental values (points) and the calculated ones (lines) follows from this figure. The values of maximum specific substrate removal rates confirm the stimulation effect of ferrous ion on activated sludge respiration activity. As it was mentioned earlier, only small differences between the values of maximum SOUR obtained for ferric iron, aluminium precipitant and control systems resulted from the work. The values of maximum SOUR given in **Table 2** also confirm the stimulation effect of iron salts on activated sludge respiration activity.

From **Table 2** is obvious that $R_{x,max}$ value for the sludge cultivated in presence of Fe^{2+} is significantly higher than the value for the sludge cultivated in presence of Al^{3+} indicating a stimulatory effect of ferric salts to the respiration rate. However, sludge cultivated in the presence of ferrous salts showed the lowest measured respiration activity. From this knowledge, it leads to the conclusion that the ferrous and ferric forms of iron have different effects on the activity of the sludge. Stimulation effect of ferric salts is highly dependent on the pH in aeration tank.

Higher stimulation of heterotrophic microorganisms activity by ferrous salts in comparison with ferric precipitant is also of technological importance, for example, with regard to the place of ferrous precipitant dosing. This form of iron salts should be preferable dosed directly into the aeration tank. In other words, ferrous sludge should not be metered prior to aerated sand traps because the presence of oxygen would oxidize ferrous iron to ferric. Ferric iron has significantly less stimulation effect on the activity of the sludge in the aeration tank as ferrous salts dispensed directly to this aeration tank. In addition, ferrous salts dosing will save operating costs for the simultaneous precipitation of phosphorus in comparison with ferric salts. However, one should keep in mind when designing activated sludge process with simultaneous phosphorus removal that dosing of the ferrous salts will increase oxygen consumption in this part of aeration tank.

The courses of the experimental and calculated values of SOUR (related to TOC content in activated sludge) at different values of substrate concentration are presented in **Figure 4**. The highest values of SOUR can be concluded for the activated sludge cultivated in the presence of ferrous precipitant. On the other hand, this figure indicates only small differences between the SOUR courses obtained for activated sludge, which was cultivated in the control system and in the presence of ferric iron or aluminium precipitate. In **Figure 5**, the similar concentration dependencies are plotted, but the values of SOUR for the same measurements are related to volatile suspended solids (VSS) concentration in each bioreactor. Similarly to **Figure 4**, the highest values of SOUR have been achieved with activated sludge cultivated in the bioreactor with ferrous precipitant dosing.

Effluent phosphorus concentration varied between 1.0 and 2.0 mg L^{-1}.

3.1.1. Nitrification tests

The aim of preliminary short-term nitrification tests was to evaluate the effect of precipitating salts on nitrification activities of individual sludge after approximately one month of sludge acclimation in presence of individual precipitating agents. Kinetic assays were performed after feeding semi-continuous models by the model substrate composed of peptone, glucose, ethanol, sodium acetate, ammonium chloride, sodium dihydrogen phosphate and of the precipitating agent. During the test, the pH was maintained at about 7.5. Measured values of different forms of nitrogen compounds during one cycle of each model are listed in **Tables 3–5**. Evaluation and comparison of nitrification rates of each activated sludge are shown in **Table 5**.

Time [h]	$N - NH_4^+$ **[mg L⁻¹]**	$N - NO_2^-$ **[mg L⁻¹]**	$N - NO_3^-$ **[mg L⁻¹]**
0	22.7	6.4	23.8
0.5	23.7	3.3	24.5
1.0	–	2.6	25.0
2.0	21.4	0.9	27.5
3.0	18.3	0.4	32.0
4.0	15.1	0.4	–

Table 3. Time variation in nitrogen compounds in control activated sludge model.

Comparing the rate of nitrification of second stage given in **Table 5**, one can conclude stimulatory effect of ferrous salts to the second stage of nitrification. The nitrification rates are based on the value of the total organic carbon (TOC) in individual models that best represent the active part of the biomass of activated sludge. The values of biomass concentrations in systems with the addition of the precipitants contain sufficient portions of the chemical sludge, and therefore, they are not suitable for expressing concentration of active biomass.

	Al^{3+}		**Fe^{2+}**		**Fe^{3+}**	
Time [h]	$N - NH_4^+$ **[mg L⁻¹]**	$N - NO_3^-$ **[mg L⁻¹]**	$N - NH_4^+$ **[mg L⁻¹]**	$N - NO_3^-$ **[mg L⁻¹]**	$N - NH_4^+$ **[mg l⁻¹]**	$N - NO_3^-$ **[mg l⁻¹]**
0	26.1	–	24.4	14.0	23.0	16.9
1	20.9	13.3	21.2	13.9	23.6	16.5
2	17.0	14.9	14.6	14.9	16.8	–
3	14,3	–	9.9	19.1	11.2	20.7
4	9.1	21.3	7.7	–	8.3	23.5
5	7.2	24.1	3.0	24.5	4.6	–
6	–	26.7	1.4	29.7	1.9	26.8
7	1.2	29.7	0.2	29.4	0.6	32.6

Table 4. Time variation in nitrogen compounds in activated sludge models with precipitating agents dosing.

	Control	Al^{3+}	Fe^{2+}	Fe^{3+}
VSS [g L^{-1}]	2.78	3.10	3.16	3.13
TOC [g L^{-1}]	0.96	1.05	0.98	0.99
r_X [mg N-NO_3 g^{-1} h^{-1}]	0.97	0.90	1.11	0.80
r_{TOC} [mg N-NO_3 g^{-1} h^{-1}]	2.80	2.67	3.58	2.53

Table 5. Nitrification rate values.

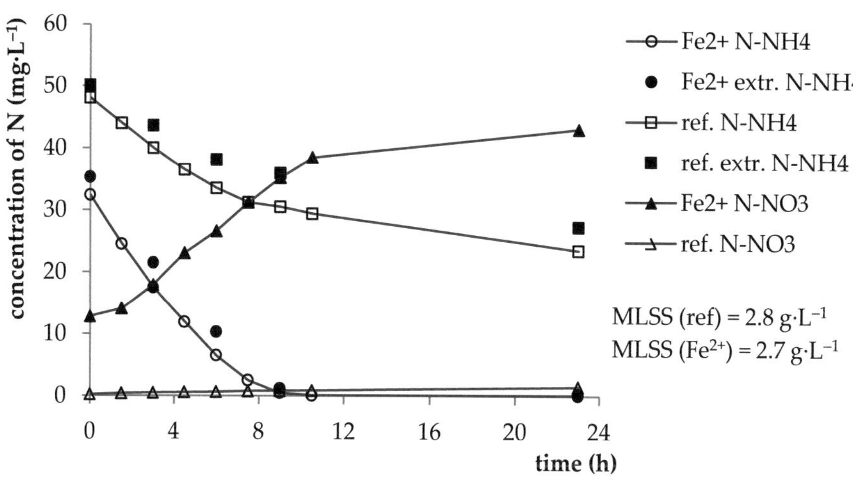

Figure 6. Time dependencies of various forms of nitrogen compounds during 24-hour cycle in the reference model and in the model with ferrous salts.

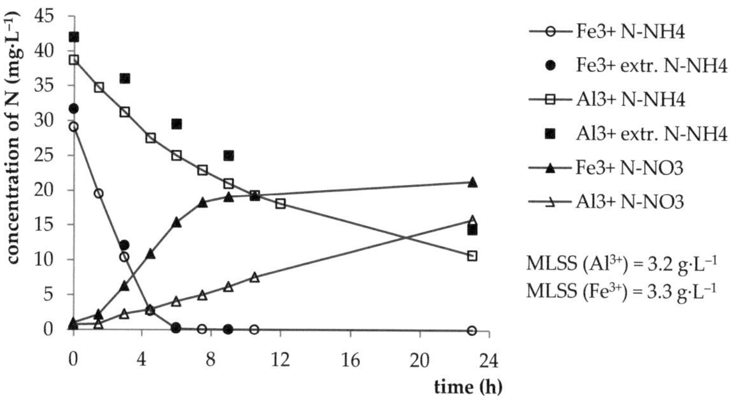

Figure 7. Time dependencies of various forms of nitrogen compounds during 24-hour cycle in the model with ferric and aluminium salts.

The rate of transformation of N-NH_4 to nitrite (1st nitrification step) cannot be taken as a benchmark to determine the impact of metal on oxidation of ammonia nitrogen since this value covers not only the rate of oxidation of N-NO_3, but also the consumption of N-NH_4 for the synthesis of biomass. Moreover, it is also influenced by the formation as a result of hydrolysis and ammonification of organically bound nitrogen.

The next kinetic tests were aimed at the investigation of nitrification during the whole reaction cycle in individual laboratory semi-continuous models. The results of the measurements are shown in **Figures 6** and 7.

In order to compare individual models, the rates of ammonium nitrogen removal in the first hours of the test (rate includes the process of assimilation and nitrification) were calculated from the experimental data. The measured rates are shown in **Table 6**. In order to thoroughly verify coagulants influence on the process of nitrogen removal, the process of ammonium nitrogen adsorption on biomass was performed. The results of sorption of ammonia nitrogen are shown in **Figures 6** and 7. From the results follows approximately equal and constant sorption of ammonium nitrogen during the whole cycle in all models. The value of adsorbed ammonia nitrogen is actually the difference between value of nitrogen extraction and the value of ammonia nitrogen measured without extraction. The concentration of ammonium nitrogen adsorbed on the biomass flakes reached 15–40% of the concentration of dissolved ammonium nitrogen, which is in good agreement with published values which ranged from 18 to 30% [23]. These values of sorption of ammonia nitrogen are characteristic for the semi-continuous system operation laboratory models.

	Control	**Al^{3+}**	**Fe^{2+}**	**Fe^{3+}**
Maximal rate of assimilation and nitrification $R_{x,N,m}$ [mg g^{-1} h^{-1}]	0.92	0.78	1.69	1.77
Rate of simultaneous nitrification $R_{x,N,m}$ [mg g^{-1} h^{-1}]	0.05	0.23	0.99	0.89

Table 6. The measured rate of assimilation and simultaneous nitrification during the daily cycle.

3.1.2. Denitrification test

The aim of the denitrification test was to monitor the degradation rate of nitrate nitrogen in individual models. For the better approximation of the active biomass, the denitrification rates were based on the concentrations of total organic carbon (TOC) in the individual sludge. The measured concentrations of nitrate and nitrite nitrogen during the denitrification tests are shown in **Table 7**. Experimental values of total suspended solids (TSS) and TOC are shown in **Table 8**.

Experimental and calculated dentrification rates based on TSS and TOC unit values in individual operated lab-scale models are shown in **Table 9**. From these results follow the stimulatory effects of both forms of iron salts, that is, ferrous and ferric salts on denitrification rate. However, about two times higher denitrification rate follows for ferric salts in comparison with ferrous salts when comparing specific denitrification rates based on TOC values. The experimental values of denitrification rates are comparable with those published in the literature, that is, 5–20 mg g^{-1} h^{-1} (based on unit mass of TSS).

	Control		Al^{3+}		Fe^{2+}		Fe^{3+}	
t (h)	$N\text{-}NO_2^-$ (mg L^{-1})	$N\text{-}NO_3^-$ (mg L^{-1})	$N\text{-}NO_2^-$ (mg L^{-1})	$N\text{-}NO_3^-$ (mg L^{-1})	$N\text{-}NO_2^-$ (mg L^{-1})	$N\text{-}NO_3^-$ (mg L^{-1})	$N\text{-}NO_2^-$ (mg L^{-1})	$N\text{-}NO_3^-$ (mg L^{-1})
0	4.4	49.0	5.2	43.9	1.6	46.1	2.7	46.1
1	5.9	42.1	8.2	34.3	7.1	35.2	6.2	18.5
2	7.3	33.0	15.4	23.2	13.7	26.2	13.7	6.6
3	7.3	21.5	17.6	15.7	15.3	10.2	14.6	6.0
4	–	16.7	19.5	10.9	19.3	5.5	19.3	5.5
5, 6	–	15.1	19.5	3.0	19.3	3.9	19.3	3.2

Table 7. Concentration values of $N\text{-}NO_2^-$ and $N\text{-}NO_3^-$ in individual models.

	Control	Al^{3+}	Fe^{2+}	Fe^{3+}
X (g L^{-1})	1.28	1.36	1.32	1.32
TOC (%)	34.5	33.75	31.02	31.62

Table 8. Concentration values of TSS and TOC in individual models.

	Control	Al^{3+}	Fe^{2+}	Fe^{3+}
$R_{X,N\text{-}NO3}$ (mg/g h)	7.2	6.9	9.1	15.0
$R_{TOC,\ N\text{-}NO3}$ (mg/g h)	20.9	28.4	29.3	47.4

Table 9. Denitrification rates in individual models.

3.2. Impact on sedimentation, thickening and dewatering characteristic

Table 10 shows experimental values of dry mass concentration for each lab model and calculated values of u_0, c_k and n parameters in the particles mass flow density Eq. (4). The minimum thickening area values are also given for individual sludge determined from point of view of maximum hydraulic surface load or establishment of conditions for sludge separation in sedimentation tank [Eq. (10)]. Eq. (5) was applied to calculate thickening rate values corresponding to experimental values of sludge concentration in each model (**Table 10**). The same value of activated sludge flow rate ($Q_0 = 250$ m^3 h^{-1}) was applied to calculate minimum thickening area for each operational activated sludge.

The smallest value of minimum thickening area results from **Table 1** for aerated tank with Fe^{3+} doses; the same area is larger by approximately 16% for reference activated model sludge (without precipitation agents). Minimum thickening area for Fe^{2+} is larger by app. 19% than in the reference model. The least thickening rate or the largest minimum thickening area are characteristic for sludge with Al^{3+} doses; this area is almost doubled in size compared to the sludge with Fe^{2+} dosing. It results from the above that this sludge shows the best sedimentation characteristics despite the highest solids concentration in the system with Fe^{3+} dosing (thickening rate decreases with growing sludge concentration).

Model	X_A [kg m^{-3}]	u_0 [m h^{-1}]	c_k [kg m^{-3}]	n [–]	u_z [m h^{-1}]	A_{min} [m^2]
References	2.5	19.2	12.2	1.58	13.3	18.8
Fe^{2+}	3.0	16.9	82.6	11.14	11.2	22.3
Fe^{3+}	3.4	24.8	88.1	12.00	15.5	16.2
Al^{3+}	3.3	10.0	77.3	13.15	5.7	44.0

Table 10. Values of dry mass concentration, parameters of particles mass flow density equation [Eq. (1)] and minimum thickening area for each sludge [Eq. (7)].

Table 11 shows experimental values of volatile suspended solids (VSS) in each lab model sludge, calculated values of *k* and ß parameters in thickening rate equations [Eq. (6)] and relevant values of minimum thickening area [Eq. (10)]. The same above specified activation mixture flow rate was used to calculate minimum thickening area for individual sludge.

Model	$X_{SŽ}$ [kg m^{-3}]	k [m h^{-1}]	β [m^3 kg^{-1}]	u_z [m h^{-1}]	A_{min} [m^2]
References	2.20	27.6	2.50	14.6	17.1
Fe^{2+}	2.24	21.1	0.16	12.9	19.4
Fe^{3+}	2.49	24.8	1.41	15.3	16.3
Al^{3+}	2.61	9.9	0.18	5.6	44.8

Table 11. VSS concentrations, thickening rate equations [Eq. (3)] parameters and minimum thickening area [Eq. (7)].

Minimum thickening area values calculated with the simpler two parameters model [Eq. (6), **Table 11**] also prove the above evaluation of sedimentation characteristics of each sludge. **Table 12** shows minimum particles mass flow density values and the necessary thickening area (sedimentation tank) calculated with both above presented mathematical models for individual sludge with dosed precipitation agent. Values of $q_{min,5}$ [Eq. (8)] and A_{Z1} [Eq. (10)] correspond to the three parameters model and $q_{min,6}$ [Eq. (9)] and A_{Z3} [Eq. (10)] correspond to the two parameters model. Calculations were made for dry matter concentrations in each activation model presented in **Table 10**, and the above specified wastewater flow rate and the same value of return sludge recirculation ration.

It results from comparison of values presented in **Table 12** that the largest necessary thickening area is required for sludge with aluminium salts dosing in order to reach the required return sludge thickening to maintain the above specified activated sludge concentrations in each system (**Table 10**) at the same flow rate condition (wastewater flow rate and return sludge

http://dx.doi.org/10.5772/65492

recirculation ratio), while five to ten times smaller thickening area corresponds to sludge with iron salts dosing. The values of required thickening area for sludge with Fe^{2+} and Fe^{3+} dosing calculated with individual mathematical thickening models [Eqs. (4) and (6)] are contradicting. Values higher by app. 23 up to 70% result for simplified, two parameters model from comparison of the measured and calculated residual variation values of particles mass flow density figures. Thus, three parameters model was better in describing experimental values [Eq. (4)].

Model	$q_{min,5}$ [kg m^{-3} h^{-1}]	$A_{Z,1}$ [m^2]	$q_{min,6}$ [kg m^{-3} h^{-1}]	$A_{Z,3}$ [m^2]
Fe^{2+}	30.1	27.5	36.2	22.8
Fe^{3+}	24.1	38.9	52.7	17.8
Al^{3+}	2.7	331.0	9.5	94.5

Table 12. Values of q_{min} a A_Z for sludge with dosed precipitation agents.

The relations among thickening rate of sludge from models with precipitation agents dosing are also obvious from **Figure 8** that shows the particles mass (MLSS/TSS) flow density curves. The highest particles mass flow density value corresponds to the selected sludge concentration and thus, also the largest thickening sludge rate with Fe^{3+} dosing. The least values of these parameters are characteristic for sludge with Al^{3+} dosing. Alongside the parameters of the applied mathematical model, the value of particles mass flow density depends also on the requested sludge thickening, which relates to sludge concentration maintained in the activation tank and also to the return sludge recirculation ratio. This may also explain the seemingly contradicting calculated values of required thickening area for sludge with Fe^{3+} and Fe^{3+} (compared to the course of particles mass flow density curves depicted in **Figure 8**). Higher sludge concentration with Fe^{3+} dosing compared to Fe^{3+}, at the same recirculation ratio, results also in higher required thickening, which was also reflected in smaller minimum particles mass flow density value or higher value of the required thickening area.

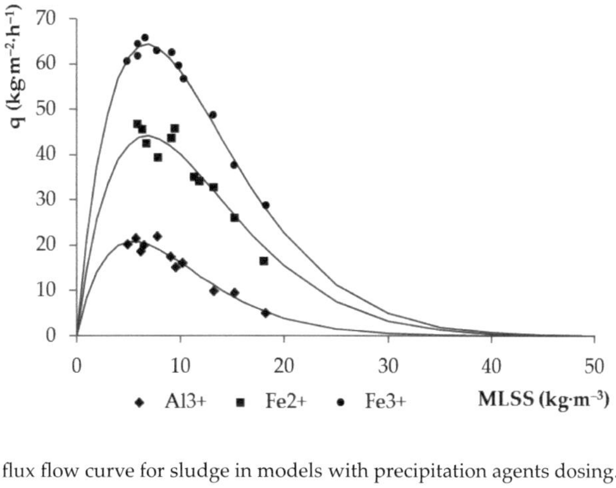

Figure 8. Particles solids flux flow curve for sludge in models with precipitation agents dosing.

Model	X_R [kg m^{-3}]	X_K [kg m^{-3}]	A_Z [m^2]
References	12.2	12.2	139.0
Fe^{2+}	32.0	27.6	135.0
Fe^{3+}	45.2	40.8	154.0
Al^{3+}	42.8	38.3	151.0

Table 13. Values of A_Z corresponding to maximum surface load in undissolved substances pursuant to.

Table 13 presents return sludge concentration X_R for each sludge required to maintain the above specified sludge concentrations in the system (**Table 10**), critical concentration X_K and the necessary minimum thickening area corresponding to the requirements of maximum surface load in undissolved substances (mass surface load) in compliance with the standard STN 75 6401 [24], that is, 6 kg m^{-3}h^{-1}. Respective values of sedimentation rates are given in **Table 13**.

It is obvious that activated sludge process intensification with the purpose of chemical phosphorus precipitation is accompanied with its increased concentration in the system, when unchanged sludge age is maintained. Thus, mass surface load may occur with insufficient sedimentation tanks dimension. It is obvious from **Tables 1** and **3** that the higher sludge concentration in the system, the higher the required thickness of returned sludge, and thus, also higher thickening area.

Figure 9 shows particles mass flow density curve for reference sludge (no agents dosed). The below sludge volume index (SVI) values were measured in individual bioreactors:

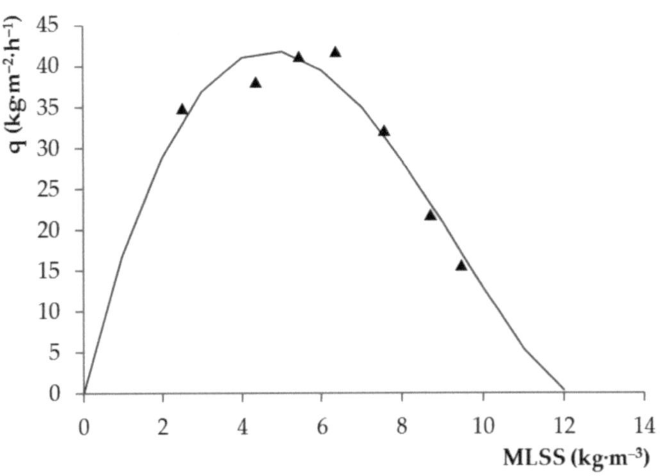

Figure 9. Particles mass flow density curve for reference sludge.

Fe^{2+}: 29.8 mL g^{-1}, Fe^{3+}: 36.3 mL g^{-1}, Al^{3+}: 42.5 mL g^{-1}and Ref: 69.9 mL g^{-1}.

These figures show high sedimentation characteristic of each sludge. SVI, however, is not a good parameter to compare sedimentation and thickening characteristics of these different sludge due to the different dry mass concentrations in each sludge, mineral portion and morphology of floccules.

Table 14 presents filtration characteristics for each sludge gained by processing filtrate volume increment measurements in individual time intervals, that is, applying Eqs. (12) and (13). The highest specific filtration resistance value for sludge cultivated with additional iron salts is obvious from **Table 14**. The lease values of fictive volume V_e correspond to this value, which also proves the least transitivity of the sludge cake. Relatively higher specific filtration resistance value and lower throughput are characteristic also for sludge with Al^{3+} dosing. The highest throughput results for reference sludge and the least specific filtration value are characteristic for sludge with addition of iron salts.

	α [m kg^{-1}]	k [s m^{-6}]	V_e [m^3]
References	0.929×10^{12}	3.05×10^{-11}	129×10^{-6}
Al^{3+}	9.750×10^{12}	0.48×10^{-11}	18.9×10^{-6}
Fe^{2+}	16.900×10^{12}	0.26×10^{-11}	0.6×10^{-6}
Fe^{3+}	0.319×10^{12}	$15{,}30 \times 10^{-11}$	26.0×10^{-6}

Table 14. Filtration characteristics for individual excess sludge measurements at $\Delta p = 19.6$ kPa—Eq. (12).

Table 15 presents specific filtration resistance values corresponding to measurement of filtrate volumetric flow rate through filtration material, filtration material and cake and filtrate volume corresponding to the developed filtration cake. Measurements were made at the same pressure difference as in the previous case, that is, $\Delta p = 19.6$ kPa. Measured data were evaluated with Eq. (17). Relative errors [Eq. (15)] did not exceed 4%. The lowest specific filtration resistance values result for sludge with dosed iron salts also from these results. When the same method is applied to define the specific filtration resistance values for other sludge used in our research, the results are very much the same while exceeding the sludge with Fe^{3+} dosing by approximately one order.

	α [m kg^{-1}]	Relative deviation [%]
References	4.56×10^{12}	0.13
Al^{3+}	4.16×10^{12}	0.55
Fe^{2+}	5.18×10^{12}	0.43
Fe^{3+}	0.63×10^{12}	3.95

Table 15. Filtration characteristics of sludge measured at $\Delta p = 19.6$ kPa—Eq. (14).

	α_0	A	x	S^2R
Control, (16)	2.552×10^{12}	0.426×10^{12}	0.0005	2.457×10^{25}
Control, (17)	–	2.958×10^{12}	0.0005	2.457×10^{25}
Al^{3+}, (16)	0.051×10^{12}	0.310×10^{12}	0.2048	0.099×10^{25}
Al^{3+}, (17)	–	0.554×10^{12}	0.1941	0.103×10^{25}
Fe^{2+}, (16)	0	0.005×10^{12}	0.6878	0.192×10^{25}
Fe^{2+}, (17)	–	0.005×10^{12}	0.6878	0.192×10^{25}
Fe^{3+}, (16)	0.008×10^{12}	0.106×10^{12}	0.2500	0.034×10^{25}
Fe^{3+}, (17)	–	0.001×10^{12}	0.7603	0.009×10^{25}

Note: numbers 16, and 17 mean that Eq. (16), or Eq. (17) were applied.

Table 16. Parameters of specific filtration resistance dependence from pressure difference Eqs. (16) and (17).

Table 16 presents the values of specific filtration resistance dependence from pressure difference, that is, the parameters of Eqs. (16) and (17). The data for calculation of specific filtration resistance values at different pressure differences were obtained by measuring volumetric flow rates through filtration material, filtration material and cake and also the filtrate volume when filtration cake was developed. Relative errors calculated with Eq. (15) did not exceed 4% for these measurements. The above mentioned equation describes the experimental data also for reference sludge and sludge containing aluminium precipitates, with approximately the same quality. The value of α_0 equals zero for sludge from model where iron salts was dosed. Better agreement of experimental and calculated α_0 values was gained with equation using zero value of α_0 also when iron salts were dosed, that is, Eq. (17).

The highest α_0 value corresponds to sludge from reference model (lower by app. two orders compared to sludge with Fe^{3+} or Al^{3+} dosing). Very low values of x parameter result for activated sludge from reference model or low specific filtrate resistance dependence on pressure difference.

4. Influence on anaerobic sludge digestion

Chemical precipitation of phosphorus produces metals' precipitates (Fe, Al). These are transported with the waste sludge to the digestion tanks where they occur at relatively high concentrations. The Fe precipitation's influence on the sludge anaerobic stabilisation process as well as on the sludge water quality was tested in this part of the work.

The Fe concentration increased in the digestion tank to app. 1000 mg L^{-1}. This, as a result of the simultaneous P precipitation in the activation, partially inhibited the CH_4 generation. When the load in the digestion tank reached 1.9 kg m^{-3} d^{-1} (kg sludge TS), the inhibition was 15–50%, depending on the number of the sludge inputs during the day. When the load decreased to 1.6 kg m^{-3} d^{-1}, the inhibition decreased to 5–20% depending on the number of sludge additions. The concentration of NH_4 and volatile fatty acids (VFA) in the digestion tank with Fe increased probably because the hydrolysis was stimulated and methanogenesis inhibited. Fe and P concentrations in the supernatant (sludge water) were minimal. On the other hand, the stimulation effect on the process of anaerobic stabilisation was measured when applying lower concentration (below 400 mg L^{-1}) of the metal salts. These phenomena can be the consequence of the stimulation of hydrolysis as a first step of organic matter degradation or the elimination of sulphides by Fe precipitation.

The most important characterisation of situation in operated reactors with the anaerobic sludge stabilisation in relation to added Fe precipitates ($FePO_4$) wasted from lab-scale aerobic models with simultaneous phosphorus removal. Anaerobic digestion reactor under current conditions contains only trace Fe concentrations from the wastewater. However, anaerobic stabilisation reactor is subsequently filled with Fe precipitates due to the application of precipitation. Lower part of **Figure 10** shows the course of Fe concentration increase in the reactor during the experiments. Calculated courses predict the experimental Fe concentrations (calculation was done based on the Fe mass balance). Two breaks on the curve represent periods when the reactor was turned off, and no sludge was added to the reactor**.** The ratio of produced CH_4 volume in reactors with and without Fe feeding clearly shows that during the first phase of the experiment the biogas production was stimulated by added Fe. Since Fe concentration reached app. 400 mg L^{-1} (related to reactor volume) inhibition of the process was observed. During the interruption of the reactor's feeding, both long-term (18th–33rd day) or short-term (57th–60st day) biomass regeneration occurred. When the sludge was fed again, very significant stimulation of methanogenesis was observed. Within several days, the process was inhibited again. The inhibition is shown in details in **Figure 10** where are set daily produced amount of CH_4.

Relatively high differences in daily biogas volume production can be explained by the fact that digestion reactors were operated with small amounts of a real raw sludge for which the differences in organic content occur. For characterisation of Fe influence on methanogenesis, the most important is the top part of **Figure 10** showing the ratios of biogas volumes produced. After 50 days of experiment at Fe concentration over 800 mg L^{-1}, (reactor volume) the inhibition reached almost level of 50% compared with control model. During this period, that is, from the beginning of experiment to its 75th day, the digestion reactors were fed once a day. During following days, the feeding procedure was changed in order to eliminate significant level of inhibition. The sludge was added in the same way as at real WWTP, that is, more times a day (3 times a day). It is recommended to add raw sludge more than 2 times a day [25].

From **Figure 11** follows that the decrease in methanogenesis rate was related also to COD accumulation in the sludge supernatant, while the analysis shown preferable presence of acetate. The facts that the decrease in methanogenesis rate increased the COD of the sludge water and that the sludge regeneration stimulated the process indicate that not only the Fe but probably also the presence of volatile fatty acids (VFA) caused the inhibition**.** The accumulation of VFA could result not only due to the fact that slower methanogenesis did not remove all VFA, but also due to the fact that the Fe presence probably significantly increased the rate of hydrolysis and acidogenesis, which caused COD accumulation. Increased concentration of NH_4–N concentration can be also considered the factor of methanogenesis inhibition. From **Figure 11**, it can be seen that inhibition of methanogenesis rate, that is, the amount of removed COD/sludge is higher in the system without Fe.

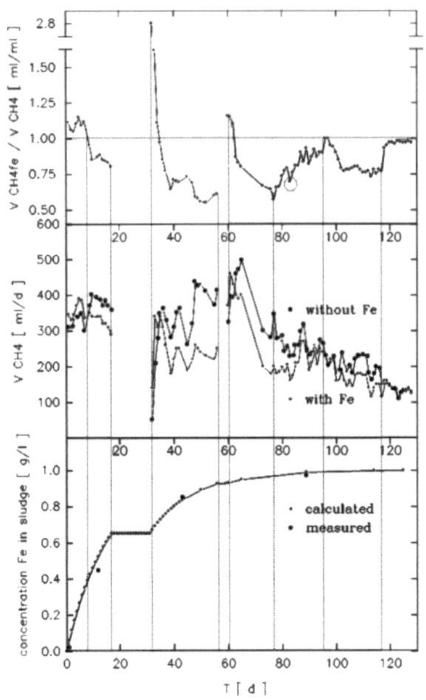

Figure 10. Time dependencies of concentration of iron precipitates and their influence on methane production during anaerobic sludge digestion.

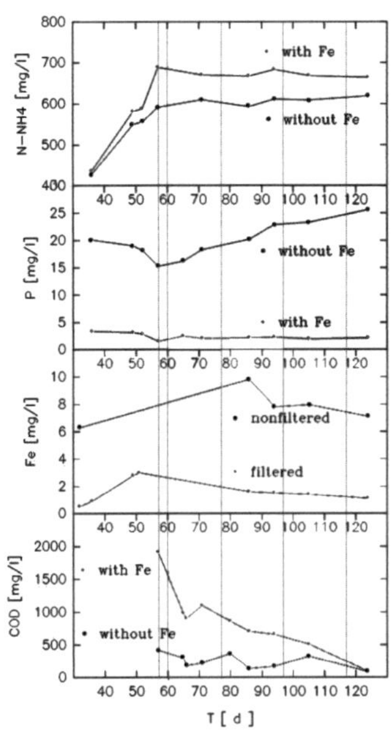

Figure 11. Time dependencies of concentration of ammonium nitrogen, phosphorus, iron and COD in sludge water during anaerobic sludge digestion.

5. Conclusions

Commonly used approaches to evaluate activated sludge activity through SOUR or specific substrate removal rate related to TSS are not fully satisfactory due to the presence of precipitates in activated sludge.

The content of VSS is a more representative quantity to relate the values of the specific rates performed by activated sludge from biological reactors operated with simultaneous chemical precipitation of phosphorus. The evaluation of respirometric measurements related to TOC content in activated sludge has proven to be the most convenient method of quantification of the effect of precipitant metals on activated sludge activity.

The decrease in SOUR of activated sludge microorganisms was observed at the initial stage of ferrous salts application. The maximum inhibition impact ranged from 2 to 6 h after the metal dosing. After the operation period of about 10 days, the values of SOUR increased. Stimulation of SOUR was observed at ferrous salts dosing. Only a small differences between the SOUR values with ferric or aluminium precipitants dosing and with sludge from control system were observed.

The highest efficiency of phosphorus removal was achieved in the bioreactor with aluminium precipitant. However, application of ferrous iron precipitant seems to be a more convenient technology of simultaneous precipitation of phosphorus due to the stimulation effect on activated sludge microorganisms activity.

Stimulation effects of ferric/ferrous salts on nitrification processes are evident. Stimulation of denitrification rates was observed for all systems with the higher stimulation in the ferric system.

The influence of dosed Fe^{2+}, Fe^{3+} and Al^{3+} salts on the sedimentation, thickening and dewatering characteristics of activated sludge was monitored with simultaneous precipitation of excess phosphorus. In addition to phosphorus elimination and activated sludge activity increase, also the improvement of sludge thickening and dewatering properties was observed.

Better description of experimental results was obtained with three parameters model of particles mass flow density curve. Comparison of minimum sedimentation tank size necessary for gravitational separation of each sludge was used to examine sedimentation characteristics of activated sludge.

Sedimentation and thickening rates increase when appropriate structure of activated sludge floccules is reached due to the higher mineral portion. Metals precipitates act as enhancement thickening agents when compared with control activated sludge thickening. The best sedimentation/thickening characteristics were reached for sludge from bioreactor where Fe^{3+} salts were dosed, while the worst sedimentation characteristics were observed in sludge with Al^{3+} dosing.

Maintaining the required sludge age with simultaneous chemical precipitation of phosphorus and unchanged volume of activation tank is accompanied with higher sludge concentration in activation, which results in higher values of particles mass flow to sedimentation tank.

Application of iron salts seems to be the best option for intensification of biological stage in existing WWTP with chemical phosphorus precipitation. Problems with creating conditions for gravitational sludge separation may occur when aluminium salts are dosed as this sludge is characterised with small and slowly settling floccules.

Thickening characteristics of sludge were evaluated based upon thickening area needed to maintain the required sludge concentration in activation or that corresponding to maximum surface load in non-dissolved substances.

The highest specific filtration resistance value was measured for activated sludge cultivated with iron salts doses, which is characterised also with the least throughput. The least value of specific filtration resistance was measured for sludge with iron salts dosing. Reference sludge was characterised by high throughput and low dependence of specific filtration resistance.

Precipitates in anaerobic digestion reactor resulted in partial inhibition of the methane generation. The concentrations of ammonium and volatile fatty acids increased, hydrolysis was stimulated, and methanogenic process was inhibited. The stimulation effect was measured for iron salts concentration lower than 400 mg L^{-1}. The feeding procedure of raw sludge into digestion reactors three times a day resulted from performed lab-scale operation research on chemical precipitation of phosphorus impact on wastewater and sludge treatment processes as the most convenient one to eliminate significant level of sludge digestion process inhibition.

Acknowledgements

This work was supported by the Slovak Research and Development Agency under the contract no. APVV-0656-12. The authors would like to thank also for the support from the VEGA Grant 1/0859/14.

Author details

Ján Derco*, Rastislav Kuffa, Barbora Urminská, Jozef Dudáš and Jana Kušnierová

*Address all correspondence to: jan.derco@stuba.sk

Institute of Chemical and Environmental Engineering, Faculty of Chemical and Food Technology, Slovak University of Technology, Bratislava, Slovak Republic

References

[1] Metcalf & Eddy Inc, Tchobanoglous G, Burton LF, Stensel HD. Wastewater engineering: treatment and reuse. 4th edition. Singapore: McGraw-Hill; 2004. p. 1408

[2] Kayser R. Wastewater treatment with nitrogen and phosphorus removal. Water distribution and wastewater engineering manual. 3rd edition. Essen: Vulkan Verlag; 1989 (in German)

[3] Bever J, Teichman H. Advanced wastewater treatment. Munich, Vienna: R. Oldenbourg Verlag; 1990. Pp. 321–322 (in German)

[4] Drtil M, Bodík I. Chemical precipitation of phosphorus. Water Manag. 1993; 9:7–10 (In Slovak)

[5] Mowat A. Measurement of Metal Toxicity by biochemical oxygen demand. J. Water Pollut. Control Fed. 1976; 48:853–867

[6] Wyttenbach A, Furrer V, Tobler L. Aluminium concentration of spruce needles. J. Trace Microprobe Tech. 1996; 14:183–190

[7] APHA, AWWA, WEF. Standard methods for the examination of water and wastewater. A. E. Greenberg, L. S. Clesceri, A. Eaton (Eds.). 22nd edition. Washington, DC: APHA; 2005. Pp. 2:71–80; 5:10–18.

[8] Čech JS, Chudoba J, Grau P. Determination of kinetic constants of activated sludge microorganisms. Water Sci. Tech. 1985; 17(2/3):259–272

[9] Čech, JS, Farkač J, Chudoba J. Simple respirometric method for estimation of kinetic constant of activated sludge microorganisms. Vodní hospodářství. 1984; B34:215–221 (in Czech)

[10] Spanjers H, Vanrolleghem PA, Olsson G, Dold PL. Respirometry in control of the activated sludge process: Principles. Scientific and technical report/International Association on Water Quality 7. London: IAWQ; 1998.

[11] Dang S, Harvey DM, Jobbagy A, Grady CPL Jr. Evaluation of biodegradation kinetics with respirometric data. Res. J. WPCF. 1989; 61(11/12):1711–1721

[12] Derco J, Drtil M, Bodík I, Hutňan M. Technologies for water resources protection. Internet Distance Education, Grant of Open Society Foundation. Nitra, Bratislava: UVTIP; 2004. p. 533 (In Slovak)

[13] Derco J, Valičková M, Šilhárová K, Dudáš J, Luptáková A. Removal of selected chlorinated micropollutants by ozonation. Chem. Papers. 2013; 67(12):1585–1593. doi: 10.2478/s11696-013-0324-x

[14] Schroeder ED. Water and wastewater treatment. New York: McGraw-Hill, Inc.; 1977. p. 370

[15] Pitter P, Tuček F, Chudoba J, Žáček L. Laboratory Methods in Water Technology. Praha: SNTL; 1983 (in Czech)

[16] Tuček F, Koníček Z. Calculation of secondary settling tank area and degree of thickening in the system activation tank – secondary settling tank. Vodní hospodářství. 1989; 6:152–158 (in Czech).

[17] Tuček F, Rešetka D. Determination of specific filtration resistance. Vodní hospodářství. 1970; 4:96–99 (in Czech)

[18] Kossaczký E, Surový J. Chemical engineering I. Bratislava: Slovak University of Technology; 1972 (in Slovak)

[19] Tuček F, Chudoba J, Koníček Z. Basic processes and calculations in water technology. Praha: SNL; 1988 (in Czech)

[20] Yoshioka N, Hotta Y, Tanaka S, Naito S, Tsugami S. Continuous thickening of homogeneous flocculated slurries. Kagaku Kogaku. 1957; 21(2):66

[21] Derco J, Zarnovský L, Kuffa R, Liptáková E. Influence of iron and aluminium salts on activated sludge process during simultaneous precipitation of phosphorus. Pol. J. Environ. Stud. 1996; 3:23–28

[22] Derco J, Kuffa R, Kušnierová J, Fargašová A, Influence of phosphorus precipitants on activated sludge activity. J. Trace Microprobe Tech. 1998; 16(4):417–425

[23] Nielsen P. H., Adsorption of ammonium to activated sludge. Water Res. 1996; 30:762–764

[24] STN 75 6401. Sewage treatment plants for more than 500 population equivalents. Slovak Standards Institute, Office of Standards, Metrology and Testing of the Slovak Republic; 1999 (in Slovak)

[25] Spinoza L, Techniques and experiences in sewage sludge management. Proceedings Conference Water Protection and Waste Water Treatment, No.11, Portoroz; 1993.